Postharvest Technology of Horticultural Crops

Postharvest Technology of Horticultural Crops

Priya Awasthi
Professor & Head (Department of Post Harvest Technology)
Banda University of Agriculture & Technology, Banda - 210001

Subhash Chandra Singh
Associate Professor (Department of Fruit Science)
Banda University of Agriculture & Technology, Banda - 210001

Rohit Kumar
Department of Post Harvest Technology
Banda University of Agriculture & Technology, Banda - 210001

-EPH-

Elite Publishing House

First published 2024
by CRC Press
4 Park Square, Milton Park, Abingdon, Oxon, OX14 4RN

and by CRC Press
2385 NW Executive Center Drive, Suite 320, Boca Raton FL 33431

CRC Press is an imprint of Informa UK Limited

© 2024 Elite Publishing House

The right of Priya Awasthi, Subhash Chandra Singh and Rohit Kumar to be identified as author(s) of this work has been asserted in accordance with sections 77 and 78 of the Copyright, Designs and Patents Act 1988.

British Library Cataloguing-in-Publication Data
A catalogue record for this book is available from the British Library

Print edition not for sale in India.

ISBN13: 9781032627625 (hbk)
ISBN13: 9781032627632 (pbk)
ISBN13: 9781032627656 (ebk)

DOI: DOI: 10.4324/9781032627656

Typeset in Adobe Caslon Pro
by Elite Publishing House, Delhi

–E P H–

Contents

Preface

India is the world's second largest producer of fruits and vegetables, however in comparison to developed countries; fewer than 2% of fruits and vegetables are processed. Food is one of the most basic needs for survival. Since prehistoric times, people have been knowledge of processing and preserving food. Recently, technical improvements have been aimed at being both cost effective and practically possible.

Post Harvest Technology is an essential branch of agriculture, and its importance grows in the field of horticulture because horticultural crops are highly perishable in nature. Although post-harvest technology has been used throughout the world from time immemorial, the business in India is still in its infancy.

The book is intended for B.Sc., M.Sc., and Ph.D. students of Agriculture and Horticulture discipline. This book is written in light of the current situation. We have taken every effort to describe each and every post harvest technological topic as accurately as possible. The book is intended for JRF/SRF/ARS/NET/ICAR-Ph.D. and SAU Ph.D. candidates. It is also useful for preparing for interviews in the smallest amount of time.

The book will be extremely beneficial to post-harvest technologists, food scientists, entrepreneurs, warehouse handlers, and teachers, undergraduate and postgraduate students from various agricultural universities.

Authors

About the Authors

Dr Priya Awasthi received her M.Sc. (Ag.) Horticulture in the year 2002 and Ph.D. in Fruit Science in the year 2006 with first class from Govind Ballabh Pant University of Agriculture & Technology, Pantnagar, Uttarakhand. She has qualified ICAR NET in the year 2004 and 2006. She has received Chancellor Silver medal in her UG programme and has been a University merit scholarship holder in UG, PG and Ph.D. programme. She has started his carrier in Jawaher Lal Nehru Krishi Vishwavidyalaya, Jabalpur as Assistant Professor. After that she has joined as Associate Professor in the Deptt. of Post-Harvest Technology, College of Horticulture, BUAT, Banda. Now she is working as Professor & Head, Department of Post-Harvest Technology. She is life member of many professional societies and recipient of Excellence in Teaching Award and Young Scientist Award from recognized professional societies. She has published 25 research papers, two books, 7 book chapters, practical manuals and more than 50 popular articles. She has participated and presented paper in many national and international seminar/conferences.

Dr Subhash Chandra Singh received M.Sc. (Ag.) Horticulture in the year 2001 and Ph.D. in Fruit Science in the year 2005 and secured first class from Govind Ballabh Pant University of Agriculture & Technology, Pantnagar, Uttarakhand. He has qualified ICAR NET in the year 2001 and 2004. He has started his carrier in Chandra Shekhar Azad University of Agriculture & Technology as Scientist-Horticulture. At present he is working as Associate Professor, Fruit Science in College of Horticulture, BUAT, Banda. He has fifteen years working experience in Bundelkhand region, U.P. He has published more than 29 scientific research papers, two books, nine book chapters, 6 technical bulletin beneficial to farmers and more than 58 popular articles. He has participated and presented paper in many national and international seminar/conferences and also recipient of 'Best KVK Scientist' and 'Excellence in Extension Work' award from recognized professional societies.

Mr Rohit Kumar is currently pursuing Ph.D. Post Harvest Technology degree from College of Horticulture, Banda University of Agriculture and Technology, Banda, Uttar Pradesh. He has obtained his M.Sc. (Horticulture) Post Harvest Technology from the same University in the year 2021 and B.Sc. Agriculture from P.B.R. Agriculture Degree College Terwa, Gausganj, Hardoi, Affiliated with Chhatrapati Shahu Ji Maharaj University, Kanpur, Uttar Pradesh. He has published 02 research papers, 02 book chapters, 02 abstracts and about 15 popular articles.

Chapter - 1

Post-Harvest Technology:
An Introduction

Due to characteristics such as crop diversity, soil, personnel, technology, geographic, and climatic circumstances, India is one of the most suited countries in the world for year-round production of fruit and vegetables. This is because of India. However, it is clear that 20% to 25% of fruit and vegetables are lost at the post-harvest stage due to poor harvesting, handling, marketing, and storage procedures. Using scientific methods for harvesting, post-harvest management, and marketing horticulture products can reduce post-harvest losses. In India, there is a tremendous demand for post-harvest technologists because of this. After harvest handling and processing, horticultural crops are the focus of postharvest technology (PHT) studies and academia. In light of the growing demand for processed products and the government's focus on reducing postharvest losses, this industry has a lot of room for growth.

INTRODUCTION

There are more fruits and vegetables grown in India than anywhere else on the planet (Anonymous, 2020a). Fruit and vegetables are a key source of income for farmers in addition to other food grains because of their short duration, high market demand, and wide range of postharvest uses. Concerns concerning postharvest losses of horticultural crops remain a pressing topic for academics in the field (Anonymous, 2012; Anonymous, 2015a). There is a specialist field of horticulture known as Postharvest Technology of Horticultural Crops, which is offered for academics and research by national and international institutes, to tackle the problem of postharvest losses in horticultural crops (Anonymous, 2017). There are many disciplines involved in postharvest management and processing of horticulture products (such as vegetables, fruits, flowers, spices, and other plantation crops), such as sorting and grading as well as cold storage and other methods.

Importance of post-harvest technology

»   Post-harvest technology is the application of technology to the post-harvest handling and storage of agricultural produce.

»   Post-harvest technology can improve the quality and shelf life of agricultural commodity.

»   Post-harvest technology can help to reduce the wastage of agricultural produce.

»   Post-harvest technology can help to improve the economics of agriculture.

»   Post-harvest technology is an important tool in the fight against hunger and malnutrition.

Objectives of Post harvest technology

»   To reduce loss in quantity or volume and the product's qualitative or nutritional value.

»   To maintain the excellent quality of the produce (color, taste, flavor, aroma).

»   To increases the shelf life of the crops or product.

»   To keep the fruits or vegetables commodities free from insects and pests.

»   To get vegetables and fruits fresh all year round.

POST HARVEST LOSSES AND RESOURCE UNDER UTILISATION IN DEVELOPING COUNTRIES

Losses in post-harvest fruits and vegetables can be difficult to forecast, and the main culprits are those caused by physiological damage and microbial interactions. According to there are numerous reasons for postharvest losses, including the following:

Food Losses after Harvesting

Mechanical damage and biological or microbiological degradation are examples of technical losses. Climate, cultural practices, poor storage conditions, and insufficient handling during shipping are only few of the factors that contribute to product degradation. Fruit and vegetable physiological deterioration is the term used to describe the natural ageing process that occurs in storage. Reactions in which intermediate and end products are unwanted are known as deterioration. As a result, the fruit or vegetable may be completely depleted of nutritional content.

Insects, bacteria, moulds, yeasts, viruses, rodents, and other animals are all examples of biological or microbiological agents. Cross-contamination by spoilage bacteria from other fruits and vegetables and containers can occur when harvested fruits and vegetables are gathered into boxes, crates, baskets, or trucks. Coryniforms, lactic acid bacteria, spore-formers, coliforms, micrococci and pseudomonas are the most common saprophytes found in fresh vegetables. Among the enterobacteriaceae, Pseudomonas and Klebsiella-Enterobacter-Serratia are the most common. Unlike bacteria, fungi such as Aureobasidium, Fusarium, and Alternaria are rather uncommon. Fruit spoiling organisms are primarily moulds and yeasts, such as *Sacharomyces cerevisiae, Aspergillus niger, Penicillum spp., Byssochlamys fulva, B. Nivea* and *Clostridium pasteurianum. Lactobacillus spp.* is also common spoilage organisms. Vegetable products can support the growth of psychotropic bacteria such as Erwinacarotovora, *P. fluorescens, P. auriginosa, P. luteola* and *Bacillus species.* Other psychotropic bacteria found in vegetables include Cytophagajhonsonae, Xantomonascampestri and Vibrio fluvialis. Both fresh and minimally processed fruits and vegetables contain these germs. Raw and processed vegetables such as cabbage, celery, raisins, fennel, watercress, leek salad, asparagus, broccoli and cauliflower lettuce juice, minimally-processed lettuce, broad-leaved and curly-leaved endive, fresh peeled hamlin oranges, and vacuum-packed potatoes are all susceptible to *Listeria monocytogenes.*

Aeromonashydrophila, a psychrotrophic and anaerobic bacterium, is a common problem in vegetables. The bacteria E. coli 0157H:7 has emerged as a major concern when it comes to food poisoning. E. coli 0157H:7 is most commonly seen in the gastrointestinal system of cattle. Because of this, feces-infected food products are a serious health concern, especially if untreated, polluted water is used to wash raw foods or swallowed directly.

Inappropriate harvesting, packaging, and transportation procedures can cause tissue wounds, abrasion, breaking, squeeze, and escape of fruits and vegetables, all of which can be exacerbated by mechanical damage. Microorganisms are more susceptible to deterioration and growth when they are damaged by mechanical forces. Even though washing can help to lessen the microbial load, it can also serve to disperse spoilage bacteria and damp surfaces to the point where they can support microbial development while the food is being stored. A wide variety of harvesting procedures produces bruising and damage to the cell and tissue structure, resulting in increased enzyme activity. Many 'after harvest losses' in underdeveloped countries are caused by transportation, handling, storage and processing. Preparing a product for sale with rough handling will result in more bruising and mechanical damage, which reduces the benefits of cooling. Due to a lack of machinery and infrastructure in poor nations, fruit and vegetable processing by-products are not fully utilized. The simplest approach to get rid of rubbish is to dump it or feed it to livestock. Animal feed and plant fertilizer compositions could benefit from waste materials like leaves and tissues. Between 49 and 80 percent of a commodity's value flows to consumers during the harvest-to-consumption process, and the rest is lost at various stages of the process.

Food Losses due to Social and Economic Reasons

Policies: This includes political constraints that make it difficult or impossible to put a technological solution into effect, such as a lack of a sound strategy to encourage and facilitate the use and management of human, economic, technical, and scientific resources to keep commodities in good shape.

Resources: Post-harvest losses can be prevented and reduced by utilizing human, financial, and technical resources.

Education: A lack of information about food preservation, processing techniques, packing, transportation and distribution is also included in this category. •

Services: As well as a lack of credit policies that address the country's demands, this relates to commercialization systems that aren't working effectively and government agencies that aren't present or working poorly.

Transportation: Fruit growers in poor nations confront a severe difficulty because the vans used to deliver bulk raw fruits to markets do not have good refrigeration equipment. Microbial degradation is accelerated when fruits are exposed to high temperatures during transit.

Pre-processing to Add Value

For crops that will be transported in refrigerated ships, land vehicles, and containers that are not built to take the entire load of field heat but can preserve pre cooled produce at a desired carriage temperature, rapid cooling following harvest is crucial.

The amount of time the commodity is expected to be stored will have a significant impact on the cooling technology that is chosen. Commodities with a short shelf life after harvest, such as those that respire rapidly, should be refrigerated as soon as possible. Because of this, pre-cooling the fruit immediately after harvest is a way to add value to the product and keep it in a condition that is acceptable to the customer.

Before freezing, juicing, or dehydrating, blanching fruits may be used as a pre-treatment procedure. To blanch the fruit, the fruit can be placed in water, steam, or hot air for a period of 1 to 10 minutes. Blanching prevents the degradation of flavour and colour and the loss of vitamins during further processing and storage by inactivating enzyme systems.

Pre-processing to Avoid Losses

Cold or high temperatures, blanching to inactivate enzymes and microbes, curing of root vegetables to improve shelf life, and chemical preservatives to suppress pests after harvest are all examples of pre-processing methods used on fruits and vegetables. Produce that is stored in a temperature and humidity-controlled environment will last longer and rot less. In order to improve the product's colour appearance and marketability, it should be packaged in the suitable material.

Alternative Processing Methods

The preservation of fruits and vegetables in rural settings can be accomplished through a variety of ways, such as fermentation, sun drying, osmotic dehydration, and refrigerated storage. In order to get rid of enzymes and microbes, pre-processed fruits and vegetables can be scalded. When it comes to food preservation, fermenting fruits and vegetables is one of the most common methods employed in rural regions, and it doesn't necessitate the use of expensive equipment. Figure-1 depicts a general schematic of the various potential methods of processing fruits and vegetables.

Raw fruits and vegetables that have been lightly treated are frequently merely cleaned and washed to preserve them (MPRFV). Detaching foreign contaminants like twigs, stalks, dirt, sand, insects, pesticides, and fertilizer residue from fruits and vegetables and their containers is the primary goal of cleaning in the early stages of processing. Light and heavy materials are separated throughout the cleaning process using a variety of gravity, flotation, picking, screening, and dewatering methods such as these. Chlorinated water is typically used for washing. A jet of air is used to keep the MPRFV product bubbling in a bath. It is possible to remove almost all air and foreign materials from the product without damaging it because to the turbulence.

Water quality is critical when it comes to washing MPRFV items, as any contamination could arise. Washing MPRFV fruits and vegetables regulates three variables, according to Wiley R.C.:

1. Quantity of water used: 5-10 L/kg of product

2. Temperature of water: 4 °C to cool the product

3. Concentration of active chlorine: 100mg/L

Two examples of specially made equipment for washing fruits and vegetables are-

1. Rotary drums for cleaning apples, pears, peaches, potatoes, turnips, and beets; high pressure water is sprayed over the product, which never comes in contact with dirty water.

2. Wire cylinder washers for leafy vegetables; these wash spinach, lettuce, parsley and leeks with medium pressure fresh water spray.

To clean fruits and vegetables of dirt in rural regions, fresh produce could be poured into plastic containers filled with tap water. To rewash and disinfect the fruit or vegetable, the unclean water in the containers could be emptied and replaced with chlorine water. Fresh produce may be chilled if electricity was available until it was processed or supplied to stores and markets.

Hot water blanching or scalding

After being submerged in hot water (or boiling water) for 1-10 minutes to lower microbial levels, fruits, fresh vegetables, and root vegetable pieces are treated with peroxides and polyphenol oxides (PPO) activity. The amount of time needed to heat the vegetable product will vary based on its type. L. monocytogenes on celery leaves has been thermally inactivated by boiling water.

Cooling in trays

Prior to being packaged in sterile plastic bags, the food must first be chilled using perforated metal trays through which cold air is circulated.

Sulphiting

By immersing the fruit or vegetable pieces (or slices), the sodium bisulphate solution is used to prevent any unwanted colour changes and to maintain a residual concentration of 100 parts per million (ppm) in the end product during this process.

Osmotic dehydration and sun drying

Dehydration is arguably the best way to preserve fruits and vegetables in rural settings. Stainless metal trays or screens spaced 2-3 cm apart is used to sun-dry fruits or vegetables pieces, which are then sliced and served. Dried produce is packaged in the same manner as fruits slices or ground floor, in plastic bags, glass bottles or cans.

By heating and drying the fruit, osmotic dehydration and crystallization preserves fruit by reducing the concentration of sugar on the fruit's surface. Depending on the final product's quality, fruits can be dried in two ways: directly or indirectly under the sun. Enzymatic processes that cause fruit discoloration and browning can be prevented with this strategy. In this way, the high sugar content in the fruit results in a dried product that retains its colour without the use of a chemical preservative.

Fermentation

Another fruit and vegetable product preservation method is this one. Cucumbers, green tomatoes, cauliflower, onions, and cabbage are all examples of vegetables that are immersed in sodium chloride solution. Lactic acid bacteria and the Aerobatic group produce enough acid during fermentation to keep food poisoning pathogens from germinating, therefore the salt content is kept at around 12 percent by weight. The fruit pulp can be fermented into wine by adding sugar and water to a solution and inoculating it with a strain of *Saccharomyces cerevisiae*, on the other hand.

Making chutneys, pickles and wine

Cucumbers that are somewhat under ripe are selected and properly cleaned with water before brining. Fermentation takes place in circular wooden vats 2.5-4.5 m in diameter and 1.8-2.5 m deep for large pickle production. Small batches of pickles can be made from 4-5 kg of cucumbers in appropriate plastic containers. A salt solution is applied to the vats after the cucumbers have been placed there. The solution is recalculated to remove concentration gradients and maintain this concentration. If the cucumbers are low in sugar content, sugar is added to the fermentation process to keep it going." After 4-6 weeks of fermentation, the salt concentration has increased to 15%. Pickles can last for months or even years if stored correctly. The yeast scum on top of the vat must be carefully monitored to ensure that the lactic acid is not destroyed. Adding a thin layer of liquid paraffin to the pickling solution will accomplish this. To remove excess salt, the pickles are steeped in hot water, then sorted and packed into jars with vinegar when the fermentation process is complete.

Sauerkraut

In order to achieve a salt concentration of 2.5 percent (by weight), selected cabbage heads are core-shredded and fermented in tap water. First, Leuconostocmesent'iroides causes rapid gas evolution; this process is responsible for much of the pleasant flavour imparted by fermented products. After that, fermentation with Lactobacillus cucumeris produces more lactic acid, followed by a third stage involving Leuconostocpentoaceticus, which produces even more lactic acid along with acetic acid, ethyl alcohol, carbon dioxide, and mannitol after about five days at 20-24°C. Lactic acid generation of 1-2 percent signifies the end of the fermentation phase. Test this by titrating the acid with sodium hydroxide (NaOH) solution, and using phenolphthalein as the colour indicator (0.1 percent w/v). After fermentation, the product is either sealed in the tank or packaged into glass jars or canned to prevent contamination. Afterwards, it is ready to be consumed.

Making of Wine

Prior to pressing, ripe fruits are delivered to the farm and sorted, rinsed, and cut. By squeezing or pulping the fruit (oranges, grapes, etc.), the juice is extracted in rural areas (mangoes, Maracay, guava, etc.). A refractometer measures the pulp's soluble solid content in Brix. Sugar can be used to raise the percentage of soluble solids to 25% if necessary.

Clarification

Bentonite or potassium ferrocyanide or other salts are used to clarify wines before they are bottled. While these treatments are meant to minimize iron complexes, which would otherwise cause the wine to darken and become cloudy, they are often insufficient for fruit wine. Microfiltration systems and cooling the wine before or after refining are two further methods for removing impurities. To clarify wine, add white gelatine (1 g per L of wine) to the fermented fruit solution and let it stand in the refrigerator for one week. This precipitates all of the suspended solids, resulting in a clear, transparent wine that can be decanted from the container's top. Simple, but effective. Flash-pasteurized or hot-filled into bottles are the most common methods for preserving the wine's flavour after clarity (100 ppm). Sodium bisulphite (200 ppm) is then added to the fruit juice and allowed to stand for 2 to 3 hours. It is at this step that the undesirable yeast flora existing in the fruit pulp is removed, allowing the added inoculums to act freely in the fruit juice and provide the desired flavour or bouquet characteristic of fruit wines. Once the yeast has been added, the juice will be ready to serve. *Acetobacter spp.* can cause an unpleasant taste and flavour if it is not kept at a low enough temperature throughout the fermentation process. After 3-4 weeks of fermenting at 22-25°C, the process comes to a close. Blending, sweetening, flavouring, and stabilizing the wines are all part of the last stage of processing. In order to maintain the product's distinctive character, the blending procedure is used to minimize the scent and flavour of some wines. Some people love single-vineyard wines although most are made with apple wine or another low-flavor spirit. Wines can be sweetened with sugar or fruit juice, which also serves to enhance the flavour of the wine. A food-grade acid, such as citric or tartaric, may be added to wine to alter acidity in some circumstances, as well. Lemon juice can be used instead of these chemicals in many rural regions. Fermentation takes place in a big bottle (18-20 L) for rural winemaking, where the contents are combined with water before being placed in the bottle. A water-filled air-lock is inserted into a hollow cork or rubber stopper inside the bottle's mouth to keep the fermentation process anaerobic. A piece of plastic tubing and a bottle are all you need to make this.

Storage

The shelf life of sun dried and fermented fruit and vegetables can be extended by several months by storing them at room temperature or in a low-temperature refrigerator (6-12 months and beyond). Wine can be stored at room temperature in glass bottles, or it can be refrigerated. Sauerkraut and pickles, two other fermented foods, are often refrigerated at room temperature.

Chapter - 2

Harvesting of Horticultural Crops

When the crops are ready to be harvested, they are gathered from the fields. Harvesting grain or pulses with a scythe sickle, or reaper is known as reaping. Ending a growth season or crop's growing cycle is a major subject of many religious festivals during harvest time. Harvesting is the most labor-intensive task of the growing season on smaller farms with limited equipment. Harvesting is done with the most expensive and complex farm equipment, like the combine harvester, on large, mechanized farms. The term "harvesting" refers to the immediate post-harvest handling, which includes sorting, cleaning, packing, and transporting the product to a wholesale or retail market, or additional processing on the farm.

Crop maturity and expected weather conditions must be taken into account while making harvesting decisions. Crop output and quality can be impacted by a variety of factors, including frost, rain, and unexpectedly warm or cold temperatures. In order to prevent adverse weather conditions, you may want to harvest your crops earlier, but this may result in a lower yield and lower quality. While delaying harvest may provide a greater crop, it also raises the likelihood of adverse weather conditions. The timing of the harvest can be a big risk for farmers.

In agriculture, harvesting refers to the collection of plant portions of commercial value. These are some examples:

Fruits-e.g. tomatoes, peppers, apples, kiwifruits etc.

Root crops-e.g. beets, carrots etc.

Leafy vegetables- spinach and swiss chard

Bulbs- onions orgarlic

Tubers- potatoes

Stems- asparagus

Petioles- celery

Inflorescences-broccoli, cauliflower etc.

Harvesting can be done mechanically or by hand. Some crops, such as onions, potatoes, carrots, and others, can benefit from a hybridization of the two methods. Hand harvesting is made easier when dirt is mechanically loosened. The type of crop, the destination, and the land to be harvested all play a role in determining which harvest method is best. It's common practice to harvest fruits and vegetables for the fresh market by hand, although machine harvesting is more common for large-scale crops.

Mechanized harvesting has a number of advantages, the most important of which is speed and cost per ton. Crops that require only one harvest can benefit from the usage of automated harvesters. Equipment purchases necessitate careful consideration of the initial investment required, maintenance expenses, and the long period of time the equipment may be required to sit idle. Besides that, the entire harvesting procedure must be designed for mechanized harvesting, including rows, field leveling and pesticide spraying, and cultivating types that can withstand rough treatment. There should be a large volume of produce that can be handled by market preparation (grade, clean, pack) and the trade.

Crops having a long harvest period benefit greatly from hand harvesting. If ripening is expedited owing to climate change and the crop must be harvested promptly, more employees can speed up harvesting. With hand harvesting over mechanized harvesting, individuals are better able to choose food at its ideal ripeness and treat it with care. As a result, the product is of higher quality and has suffered less damage. Tender crops require this. However, the harvest crew must be properly trained and supervised.

The quality of the goods harvested is also influenced by the contracting agreements made with harvest workers. Harvesting is done with care when wages are paid weekly, biweekly, or monthly. However, harvesting can become reckless when payment is made by box, meter of row, or number of harvested plants. The quality of work is also affected by the creation of teams and the division of labour. Short breaks or long work days can lead to needless rough treatment of products, as can exceptionally adverse weather circumstances (either cold or heat). As a result of this training, harvest workers will have the skills and knowledge essential to select fruit at the right stage of ripeness or maturity in order to minimize damage.

HARVEST RIPENESS AND READINESS FOR HARVEST

Harvest ripeness and harvest readiness are often used interchangeably. However, the term "ripeness" is more precise when applied to fruits like tomato, peach, pepper etc. In this stage, the flavour, texture and colour of the food have changed. However, the term "readiness for harvest" is preferred in species where these modifications do not occur, such as asparagus, lettuce and beets. Fruits' maturity is the most commonly used harvest index. There must be a distinction between physiological maturity and commercial maturity. When development is complete, the latter occurs. If the market demands commercial maturity, the ripening procedure can be used or not. When a fruit reaches physiological maturity, it displays one or more visible indicators. It is impossible to remove the seeds from a tomato because of a viscous material that encases the inside locales. Seeds harden and the fruit's inside begins to colour in peppers. During over maturity or over ripening, the fruit softens and loses some of its distinctive taste and flavour. However, it is perfect for producing jams or sauces because of the high humidity. There is no guarantee that a company's commercial maturity will coincide with its physiological development. It's not uncommon for commercial maturity to arrive much before the conclusion of development for crops including cucumbers, zucchinis, snap beans, peas, and tiny carrots.

Climactic and non-climacteric fruits must be distinguished at this juncture in the discussion. Tomatoes, peaches, and nectarines are examples of climacteric fruits. Even if they are not attached to the mother plant, they can still produce ethylene, the hormone needed for ripening. Peppers, citrus and other non-climacteric fruits and vegetables are examples of this. On the plant, commercial maturity can only be achieved. From a ripening perspective, climatic fruits are independent, and changes in flavour, aroma, colour, and texture are connected to a fleeting respiratory peak and tightly linked to autocatalytic ethylene production. Even when they are harvested green, climatic fruits like tomatoes develop their full red colour. In contrast, non-climacteric fruits like bell peppers see small colour changes after harvest. Only when the fruit is still attached to the plant will it achieve its full crimson colour. Generally speaking, a product's post-harvest life is shorter the more developed it is. This means that climacteric fruits must be plucked as soon as possible, but always after achieving full physiological maturity for distant markets. The most visible exterior signs of ripening are colour changes. It is a result of the breakdown of chlorophyll (the green colour) and the creation of certain pigments. Chlorophyll decomposition in some fruits, such as lemons, permits the yellow pigments that are already present to be more visible. The green colour, on the other hand, obscures them. Several other fruits, such as nectarines, peaches, and apples, have more than one colour: the ground colour is related with ripeness, and the cover colour in many cases is particular to the variety. Colour charts and colorimeters can be used to determine a child's maturity level depending on the proportion of desired colour.

This is the most commonly used harvest index for vegetables and some fruits, especially those that are harvested before they are fully developed. In the case of snow peas, haricot beans and other snap beans, harvest occurs when the stems emerging from soil reach a certain length, while harvest occurs before the cotyledon of soybean, alfalfa, and other legume sprouts has expanded. In addition, harvest occurs in the case of asparagus when the stems emerging from soil reach a certain diameter. Leafy greens (such as lettuce or cabbage) are harvested for their compactness, while roots like beets and carrots are harvested for their "shoulder" breadth. Many vegetables, such as spinach, use plant size as a harvest indication. There are exceptions to this rule when it comes to root crops such as sweet potatoes and potatoes.

When a crop is ready for harvest, several of its outward indicators can be seen. Onion tops falling off, melon pedicel abscission layers developing; pumpkin epidermal hardness and nut shell fragility are a few examples. Sweet corn is picked when the kernels are plump and no longer "milky" in bananas, mangoes, and other fruit.

Most fruits and vegetables are harvested based on either colour or development stage, or even a combination of the two. However, they are frequently used in conjunction with other objective metrics. To name a few: firmness (apples, pears, stone fruits), tenderness (peas), starch (apples, pears), soluble solid content (melons, kiwifruit), oil content (avocado), juiciness (citrus), sugar content/acidity ratio (citrus), scent (certain melons), and so on. It is critical for the harvesting schedule to maintain a consistent flow of raw material for processing crops. As a result, it is common practice to compute the number of days till flowering and/or the amount of heat units that have been accumulated.

Handling during Harvest

There are numerous other fieldwork tasks involved in harvesting. This includes those with a profit motive. Pre-sorting, the elimination of foliage, and the removal of other inedible components are a few examples of activities that can help in market preparation. In some instances, the product is entirely ready for the field market. However, it is customary to transfer the harvest into larger containers before transporting them to the packinghouse. They are either dry here or have water poured onto the grading lines. While these processes are being done, bruising that has a cumulative effect may have an impact on the final product's quality. There are various kinds of lesions. Loss of tissue integrity leads to wounds (cuts and punctures). This kind of damage is common during harvest and is primarily caused by the equipment used to remove the plants. Pickers' nails or peduncles from other fruits are examples of additional causes. This allows rotting fungus and bacteria to infiltrate produce. This kind of damage is simple to spot and is typically eliminated during grading and packaging. More people experience bruises than wounds. When the product is in the consumer's hands, they are less obvious and manifest a few days later. Bruises typically have one of three causes:

Inflict: Damage brought on by the impact of fruit rubbing against other fruit or by dropping the fruit (or packed fruit) onto a hard surface. These injuries are typical of the harvest and packing processes.

Deformation caused by compression: The weight of the mass of fruits on the bottom layers is what frequently causes this to happen during bulk shipping and storage. Additionally, it can occur when the packed mass exceeds the container's volume or when fragile boxes or parcels collapse under the weight of those stacked on top of them.

Abrasion: Surface harm caused by any kind of friction (packaging materials, other fruits, packing belts, etc.) against delicate fruit like pears. Abrasion causes the removal of protective scales in onions and garlic.

The type, maturity, type, and severity of the bruise all determine the symptoms. They build up over time and, in addition to having a traumatic effect, set off a number of stress-related reactions, including the beginning of healing mechanisms. The following describes this physiological response: a short rise in respiration that is linked to degradation; a momentary synthesis of ethylene that hastens maturation and aids in softening. The creation of secondary chemicals that may impact texture, taste, appearance, scent, or nutritional value occurs when mechanically disrupting membranes brings enzymes into touch with substrates. Due to injury, cell death, and the loss of tissue integrity, firmness at the impact site rapidly declines. The severity of the harm increases with product maturity. Higher temperatures and longer storage times amplify its impact. Healing takes longer when ethylene is removed or neutralized in a controlled or modified atmosphere. However, atmospheric composition also slows down the rate at which the body responds to stress.

Harvestimg Recommendations

It is advised to harvest in the chilly early hours if the time of day may be chosen. Products are more turgid as a result. Less energy is also needed for refrigeration. The time it takes for the crop to reach harvesting maturity depends on how close the market is to the final destination.

It is necessary to keep harvested goods in the shade until they are ready for shipping. Harvesting scissors and knives need to be sharp enough to prevent tearing off while still having rounded ends to prevent punctures. Harvest containers must to have soft, smooth surfaces without sharp edges. Move field containers with caution and avoid overfilling them. When transferring goods to different containers, keep drop heights to a minimum.

Teach harvest workers how to handle produce gently and to recognize the ideal harvest maturity. To prevent fruit damage during harvest and handling, wear gloves.

Curing

In some crops, curing is a necessary step after harvesting to produce a high-quality product. It is a mechanism that causes a quick reduction of surface humidity. It also causes some tissue alterations while halting further dehydration. It also serves as a barrier to infection infiltration. Curing is the process of drying the external scales, developing the colour, and closing the neck in onions and garlic. Hardening the skin of root crops, such sweet potatoes, yams, and tubers like potatoes, reduces skinning during harvest and handling as well as the formation of the healing per derma on wounds (suberisation). When a fruit or vegetable is "cured," the skin hardens on pumpkins and other cruciferous vegetables while a layer of lignified cells naturally forms on citrus fruits. This stops pathogens from growing and forming. Curing typically takes place outside. It is done to protect plants from direct sunlight by undercutting and windrowing garlic and onion plants, or by placing them in piles or burlap bags for a week or longer. Potato tubers must stay in the soil for 10 to 15 days after herbicide use has damaged the foliage. Although it is typically done in shelters, it is comparable to what is done with sweet potatoes and other roots. If necessary, curing can be done artificially in storage facilities by forcing hot, humid air to circulate (Table 1). Temperature and relative humidity are regulated for long-term storage following cure.

Table-1. Recommended temperature and relative humidity conditions for curing

Crops	Temperature (°C)	Relative Humidity (%)
Yam	32-40	90-100
Potato	15-20	85-90
Cassava	30-40	90-95
Sweet potato	30-32	85-90
Onion & Garlic	33-45	60-75

Harvest Festivals

An annual celebration that takes place around the time of a region's primary harvest is known as a harvest festival. Harvest celebrations occur at various periods around the world due to regional variations in climate and crops. Harvest celebrations often include family and public eating with food derived from crops that reach maturity around the time of the celebration. The two main aspects of harvest celebrations are the abundance of food and the absence of the need to work in the fields. Eating, fun, competitions, music, and romance are other aspects that are present at harvest festivals all over the world. One of the most well-known harvest festivals in the world is the Chinese Moon Festival, which is celebrated in Asia. A few well-known harvest festivals in India are **Onam** in August and September, **Hali** in February and March and **Pongal** in January. In North America, the months of October and November are dedicated to Thanksgiving celebrations in both Canada and the US. The origins of many religious holidays, including Sukkot, can be found in harvest celebrations. People bring food from their gardens, allotments, or farms into churches, chapels, and schools in both Britain and Canada. The food is frequently given away to the elderly and underprivileged members of the neighbourhood, or it is sold to raise money for the church or a charitable organization.

The festival now takes place at the conclusion of the harvest, which varies across Britain. The Harvest Festival may occasionally be scheduled on different Sundays by neighboring churches so that people may attend one another's Thanksgiving services. A large meal known as a harvest supper was served as a celebration of the harvest's completion. Harvest Suppers are still held in a few churches and communities.

When the Reverend Robert Hawker hosted a special thanksgiving ceremony at his church in Morwenstow, Cornwall, in 1843, it marked the beginning of the modern British tradition of holding Harvest Festival celebrations in churches. His concept of a harvest festival and the yearly practice of decorating churches with homegrown produce for the Harvest Festival service were made more popular by Victorian hymns like "We 'plough the fields and scatter,""Come ye thankful people, come," and "All things bright and beautiful," as well as harvest hymns in Dutch and German translation.

There has been a change in emphasis in many Harvest Festival celebrations as British people have begun to rely less largely on home-grown products. Churches are increasingly connecting Harvest with knowledge of and care for those in the poor world who still struggle to produce crops of sufficient quality and quantity. During the harvest season, development and relief organizations frequently generate materials for churches to utilize that emphasise their own concerns for those in need all over the world.

Chapter - 3

Post-Harvest Handling

Postharvest handling, which includes cooling, cleaning, sorting, and packing, is the phase of crop production that comes just after harvest. A crop starts to degrade the moment it is lifted out of the ground or cut off from the parent plant. Whether a crop is sold for fresh consumption or utilized as a component in a processed food product, post-harvest treatment significantly impacts final quality.

The primary objectives of post-harvest management are to prevent physical damage, such as bruises, to prevent moisture loss and slow down unwanted chemical changes, and to keep the food cool to prevent moisture loss and postpone spoiling. Sanitation is also crucial in order to lessen the risk of pathogens being transferred from polluted washing water to fresh vegetables, for instance. Post-harvest processing is typically continued in a packing house after the field. A big, complex, mechanized facility with conveyor belts, automated sorting and packing stations, walk-in coolers, and other equipment may be used in place of the more basic shed that offers shade and running water. Processing can actually start during the actual harvesting process in mechanical harvesting, with the machinery performing initial cleaning and sorting.

To retain quality, the initial post-harvest storage conditions are crucial. Each crop has a best temperature and humidity range for storage. Additionally, certain crops cannot be stored in close proximity without risking undesired chemical interactions. Particularly in large-scale operations, a variety of high-speed cooling techniques and sophisticated refrigerated and atmosphere-controlled conditions are used to extend freshness. The fundamentals of post-harvest handling for the majority of crops are the same regardless of the amount of harvest, from a family garden to an industrialized farm: handle with care to avoid damage (cutting, crushing, bruising), chill immediately and retain in cool conditions, and cull (remove damaged items).

Fruit and vegetables begin to actively rot as soon as they are harvested. The initial makeup of the crop is continuously altered by a variety of biochemical processes (postharvest physiology) until it loses its marketability. The term "after harvest freshness" refers to the time after a harvest before significant change has taken place. Since freshness affects a product's quality in a significant way, its assessment should be based on objective standards; nevertheless, up until recently, only sensory tests or mechanical and colorimetric (optical) criteria were utilized. In a recent study, efforts were made to identify biochemical markers and fingerprinting techniques that may be used as a freshness index.

MATURITY INDEX FOR VEGETABLES AND FRUITS

The guidelines for when to harvest a fruit or vegetable at each stage of development are essential to its subsequent storage, marketable life, and quality. Fruits and vegetables go through three stages of development after harvest: maturation, ripening, and senescence. When a fruit reaches maturity, it is prepared for harvest. Even though the edible portion of the fruit or vegetable may not be quite ready for immediate ingestion at this point, its size is fully matured. Following or converging with maturation, ripening makes produce palatable, as evidenced by taste. The final stage, known as senescence, is marked by the fruit or vegetables natural degeneration, including a loss of texture, flavour, etc. The sections that follow provide descriptions of a few common maturity indices.

Skin Tone: Since the skin colour of fruit changes as it ripens or matures, this element is frequently used to fruits. Depending on the type of fruit or vegetable, some don't change colour much as they get older. Although the harvester's judgment must be used, colour guides are available for cultivars of apples, tomatoes, peaches, chilli peppers, and other fruits and vegetables.

Optical Techniques: Fruit maturity can be determined using light transmission parameters. These techniques are based on the fruit's decreased chlorophyll content during ripening. A strong light is shone upon the fruit before being turned off, leaving the fruit in complete darkness. Next, a sensor monitors the fruit's light output, which is correlated to its chlorophyll concentration and, consequently, to its maturity.

Shape: Fruit's form can change as it ages and can be used as a characteristic to gauge harvest readiness. For instance, as a banana grows on the plant, its cross-section becomes less angular and more rounded. During maturation, mangoes also undergo form change. The angle between the mango's shoulders and the location where the stalk is attached may shift as the fruit ripens on the tree. When a mango is young, its shoulders slope away from the fruit stalk; as it ages, however, the shoulders become level with the point of attachment and, in some cases, may even rise above it.

Size: The time of harvest is typically determined by changes in crop size as it grows. For instance, *Zea mays* saccharata partially mature cobs are marketed as sweet com, while even less mature and consequently smaller cobs are promoted as baby com. Individual finger widths can be used to gauge banana harvest maturity. The calliper grade is determined by placing a finger in the middle of the bunch and using callipers to measure the bunch's widest point.

Aroma: As fruits ripen, they produce volatile compounds. These substances give fruit its distinctive scent and can be used to gauge whether it is ripe or not. These doors may only be usable in commercial settings if a fruit is fully ripe, when it may only be detectable by humans.

Opening a fruit: Some fruits, like ackee tree fruit, which contains hazardous quantities of hypoglycine, may produce toxic chemicals as they ripen. When the fruit is fully developed, it splits open to reveal black seeds with yellow arils. It has been demonstrated that it currently contains either very little or no hypoglycine. The fruit will have a very short post-harvest life due to its maturity, which poses a marketing challenge. Hypoglycine 'A' (hyp.) levels in ackee tree fruit were similar to those in the arils, and the seed contained noticeable amounts of hyp. at all stages of maturation, at about 1000 ppm. This investigation confirms previous findings that ackee fruit that has not yet opened or that has only partially opened should not be consumed, whereas fruit that has naturally opened to a lobe separation of over 15 mm provides little health risk as long as the seed and membrane sections are removed. These findings concur with those of Brown et al. who recommended against ever forcing brilliant red, full-sized ackee open for human ingestion.

Leaf Alterations: When to harvest fruits and vegetables depends frequently on the condition of the leaves. In root crops, the state of the leaves can also reflect how the crop is doing underground. For instance, the best time to harvest potatoes for storage is right after the leaves and stems have perished. The skins will be more susceptible to storage illnesses and less resilient to damage from harvesting and handling if they are harvested earlier.

Abscission: An abscission layer develops in the pedicel as part of a fruit's normal development. For instance, in cantaloupe melons, picking the fruit before the abscission layer has fully formed results in less flavorful fruit than fruit that is allowed to ripen for the entire amount of time.

Firmness: A fruit's texture can alter as it ages, notably during ripening when it can quickly become softer. Crop texture may also be impacted by excessive moisture loss. The harvester may only need to gently crush the fruit in order to determine whether the crop can be harvested because these textural changes are detectable by touch. The texture of fruits and vegetables can now be measured with sophisticated tools, such as texture analyzers and pressure testers, which are currently offered in a variety of forms. The fruit's surface is forced, allowing the penetrometer's or texturometer's probe to pierce the fruit flesh and provide a reading on firmness. Because the angle at which the force is applied affects the basis on which hand held pressure testers are used to determine firmness, the findings they provide may be inconsistent. The Magness-Taylor and UC Fruit Hardness testers are two regularly used pressure testers to gauge the firmness of fruits and vegetables. Instruments like the Instron Universal Testing Machine are used in a more complex test, which isn't always more effective. When reporting test pressure data or attempting to establish standards, it is important to identify the instrument and the parameters employed.

The California Agricultural Code declares that "Bartlett pears are deemed mature if they meet one of the following criteria: (a) the soluble solids in a sample of juice from not fewer than 10 representative pears for each commercial size in any lot are not less than 13 grammes; (b) the average pressure test of not fewer than 10 representative pears for each commercial size in any lot does not exceed 23 lb (10.4 kg) "", itercent. This Code is shown in Table 1 and specifies the minimum maturity for Bartlett pears.

Table 1. Minimum maturity standard of fresh Bartlett pears for selected pear size ranges

Pear Size Minimum Soluble Solids (%)	6.0cm to 6.35cm Maximum Test Pressure (kg)	£6.35c
Below10%	8.6	9.1
10%	9.1	9.5
11%	9.3	9.8
12%	9.5	10.0

Juice Volume: As fruit ripens on the tree, the amount of juice in many fruits rises. To determine a fruit's juice content, a representative sample of the fruit is obtained, and the juice is then extracted according to a set of rules and guidelines. Juice volume is proportional to the juice's initial mass, which is a function of its maturity.

Oil content and percentage of dry matter: Fruits like avocados can have their maturity gauged by their oil content. Avocados must contain no less than 8% oil by weight, excluding skin and seed, at the time of harvest and at any time subsequently, according to California's Agricultural Code. As a result, an avocado's moisture content and oil content are connected. Weighing 5–10 g of avocado pulp and extracting the oil in a distillation column with a solvent yields the oil content (figure 1). For cultivars with high oil content naturally, this technique has been effective. The solvent is put into a flask that is spherical. During the extraction process, heat is supplied via an electric plate and water is pumped to maintain a steady temperature. Utilizing solvents such petroleum ether, benzene, diethyl ether, etc., extraction is carried out over the course of 4-6 hours. Following the extraction, the water is evaporated at 105°C in an oven to recover the oil by bringing the weight of the flask to a consistent level.

Moisture Level: The oil content of avocado fruit rises while the moisture level rapidly falls throughout fruit development. Table 2 lists the moisture requirements for a variety of avocados grown in Chile to achieve good acceptance.

Sugars: As climacteric fruits mature, starch builds up as a source of carbs. The fruit's starch breaks down into sugar as it ripens. Sugar tends to build during maturation in non-climacteric fruits. Using a brix hydrometer or a refractometer is a simple way to determine how much sugar is in fruits. The amount of soluble solids or sugar content is determined by placing a drop of fruit juice in the sample holder of the refractometer and taking a reading. Many regions of the world utilise this element to define maturity. By shining light on the fruit or vegetable and measuring the amount of light transmitted, the soluble solids content of fruit can also be ascertained. However, as this is a laboratory method, it might not be appropriate for production at the village level.

Table 2. Moisture content of avocado fruit cultivated in Chile

Cultivar Moisture content	(%)
Negra de 1a Cruz	80.1
Bacon	77.5
Zutano	80.5
Fuerte	77.9
Edranol	78.1
Hass	73.8
Gwen	78.4
Whitesell	79.1

Fiber content: A valid method for assessing the ripeness of pear cultivars is to measure the amount of starch present. The procedure entails slicing the fruit in half, then dipping the halves into a solution containing 4% potassium iodide and 1% iodine. Where starch is present, the sliced surfaces stain to a blue black colour. As harvest time draws near, sugar transforms from starch. When testing reveal that 65–70% of the sliced surfaces have become blue–black, harvesting can commence.

Acidity: The acidity of many fruits changes as they ripen and mature, and in the case of citrus and other fruits, the acidity gradually decreases as the fruit ages on the tree. An indicator of the best times to harvest these fruits can be obtained by taking samples, extracting the juice, and titrating it against a common alkaline solution. Acidity is typically measured in relation to soluble solids, yielding what is known as the brix: acid ratio, rather than as a standalone indicator of fruit maturity. In some trials in Venezuela, Sanchez et al. investigated the impact of using 2-chloroethyl phosphoric acid ("ethephon") to accelerate the maturation of banana (Musa sp (L.), AAB) "Silk" fruits. There were four different treatments (0, 1000, 3000, and 5000 ppm). The "ethephon" treatments enhanced acidity and total soluble solids, according to the data. While the pH was unaffected, the production of sucrose increased. On the other hand, as shown in Table 3, the Brix/acidity ratio increased in proportion to the "ethephon" dose.

Table 3. Effect of ethephon on the maturity index (Brix/acidity ratio) of banana (manzano) Silk" fruits.

Stageofmaturity	Dm;s	Ethephondoses (ppm)			
		0	1000	3000	5000
Green	1	29.35	23.99	20.59	19.31
Slightlyripen	3	33.27	33.53	58.29	46.27
Slightlyripen	5	51.15	66.44	63.01	57.00
Slightlyripen	7	60.69	69.35	64.31	68.35
Ripen	9	53.27	57.36	54.67	55.42
Variation(%)		81.50	139.10	165.52	187.00

Means with different letters in arrow are significantly different at $p<0.05$.

Particular gravity: Specific gravity measures the weight of a substance in relation to pure distilled water at 62°F (16.7°C), which is regarded as having a specific gravity of one. By comparing the weights of similar bulks of other bodies to the weight of water, one can determine specific gravity. In actuality, the fruit or vegetable is first weighed in clean water and then in air. The specific gravity is calculated by dividing the weight in air by the weight in water. This will guarantee a trustworthy indicator of fruit maturity. A fruit's specific gravity rises as it ages. Although it is rarely utilized in practice to identify the time of harvest, this characteristic might be employed if a proper sampling method could be developed. However, it is utilized to classify crops at the post-harvest stage according to various maturities. This is accomplished by submerging the fruit in a tank of water, with the fruit that floats being younger than the fruit that sinks.

Packaging for harvest

for employees harvesting fruits and vegetables in the field, harvesting containers must be simple to handle. Many crops are bagged after harvesting. For products like citrus fruits and avocados that have firm skins, harvesting bags with shoulder or waist slings are an option. These packaging units, which can be made of a number of materials including paper, polyethylene film, sisal, hessian, or woven polyethylene, are reasonably priced but offer little protection to the crop against handling and transit damage. For crops including potatoes, onions, cassava, and pumpkins, sacks are frequently employed. Baskets, buckets, carts, and plastic crates are further varieties of field harvest storage containers. Weaved baskets and bags are not advised for high risk products due to the possibility of contamination.

Equipment for harvesting

several tools are used to harvest produce, depending on the fruit or vegetable. Secateurs or knives, as well as picking shears that can be hand-held or mounted on a pole, are often used instruments for harvesting fruits and vegetables. Mangoes and avocados are examples of vegetables or fruits that are challenging to catch. To prevent damage to the fruit when falling from tall trees, a cushioning material is placed around the tree. For crops with firm skins, such as citrus and avocados, harvesting bags with shoulder or waist slings can be utilized. They are lightweight and free up both hands for carrying. Without tossing the bag, the contents are discharged into a field container through the bottom of the bag. Harvesting crops like tomatoes that are easily smashed is best done in plastic buckets. The product should not be damaged by these containers' sharp edges, which should be smooth. Commercial growers send commodities like apples and cabbages to huge packinghouses for selection, grading, and packing in bulk bins with a capacity of 250–500 kg.

Methods and temperatures for cooling

Produce is cooled using a variety of techniques after harvesting in order to increase shelf life and preserve a fresh-like quality. The following low temperature treatments are listed for consideration even if they are not appropriate for simple rural or village treatment:

Pre-cooling: Fruit is pre-cooled when it has been cooled to a safe transportable temperature of between 3 and 6°C (5 and 10°F). Cold air, cold water (hydro-cooling), direct contact with ice, or the evaporation of product water under a partial vacuum are all options for pre-cooling (vacuum cooling). Hydro-cooling is an invention in vegetable cooling that uses chilled air and water to create a mist.

Pre-cooling the air: The most typical method is to pre-cool fruits using cold air. It is possible to carry it out in forced air coolers, storage rooms, tunnels, and refrigerator cars (air is forced to pass through the container via baffles and pressure differences).

Icing: Produce boxes often have ice added to them by covering the crop with a layer of crushed ice. The ratios below can be used to apply ice slurry: 40 percent water, 60 percent coarsely crushed ices and 0.1 percent salt chloride to lower the melting point are the ingredients. The proportion of water to ice can range from 1:1 to 1:4.

Air conditioning: The produce is stored in a chilly environment using this technique.

Despite possible variations in the type of room used, most refrigeration units circulate cold air through fans. Air may be blasted across the top of the room by the circulation and fall through the crop by convection. Cost savings are the key benefit because no special facility is needed.

Forced air cooling: This technique of pre-cooling works on the basis of putting the crop into a space where cold air is circulated through it after passing over various refrigerated metal coils or pipes. High-velocity air blowing caused by forced air cooling systems dries out the crop. Different techniques for humidifying the cooling air have been developed to reduce this effect, such as blowing the air via cold water sprays.

Hydro-cooling: Faster than the transfer of heat from a solid to a gas is the transfer of heat from a solid to a liquid. As a result, crops can be promptly cooled using cold water and will not lose weight. The crop is immersed in cold water that is continuously pumped through a heat exchanger to produce maximum performance. A hydro-cooler may be used during the water-based transportation of crops around the pack house. The benefit of this technology is that the conveyer's speed may be changed to match the amount of time needed to cool the produce. The crop can be protected from spoiling by using chlorinated water. For vegetables including asparagus, celery, sweet corn, radishes, and carrots, hydro-cooling is frequently utilized, but it is rarely used for fruits.

Vapour cooling: In this instance, cooling is accomplished by the latent heat of vaporization rather than through conduction. Water will boil at 100°C at standard air pressure (760 mmHg). The boiling point of water decreases with decreasing air pressure, and at 4.6 mmHg, it boils at 0°C. Under these conditions, the crop loses around 1% of its weight for every 5 to 6°C drop in temperature. By misting the fruit with water either before placing it in the vacuum chamber or near the conclusion of the vacuum cooling process, this weight loss can be reduced (hydro-vacuum cooling). The ratio of the crop's bulk to its surface area influences how quickly and effectively it cools. Leaf crops like lettuce are especially suited to this strategy. Vacuum cooling is not recommended for crops with a wax cuticle that is rather thick, such as tomatoes.

Transport to the Packing House and Field Packing

When picking berries for the fresh market, machines are frequently used. The berries are then typically packed into shipping containers. To maintain product quality, fruits and vegetables must be harvested, handled, and transported carefully to packing facilities.

Plastic bags: Banana bunches are packaged in the field with clear polyethylene bags and then moved to the packinghouse via mechanical cableways that pass through the banana plantation. Bananas that are packaged and transported using this method sustain less damage from careless handling.

Plastic field boxes: Usually, polypropylene, polyethylene, or polyvinyl chlorides are used to make these kinds of boxes. They are enduring and have a long lifespan. Many may stack one on top of the other without crushing the fruit because they are made to nest within one another when empty, making transportation easier.

Field boxes made of wood: These boxes are constructed from skinny pieces of wood that are wire-bound. The bushel box, which has a volume of 2200 in3 (36052 cm3), and the half-bushel box are the two sizes available. They have the advantage of being affordable and flat packable, which makes them potentially non-returnable.

Bulk bins: Fresh fruit and vegetable picking takes place in bulk bins with a 200–500 kilograms capacity. In terms of fruit carried per unit volume, durability, and offering greater product protection while transport to the packinghouse, these bins is significantly more affordable than the field boxes. They are constructed from plastic and wood materials. These bins are 48 × 40 inches in size in the United States and 120 x 100 centimeters in nations that use the metric system. Depending on the kind of fruit or vegetable being delivered, bulk bins might vary in depth (Table 4)

Table 4. Approximated depth of bulk bins

Commodity	Depth (cm)
Citrus	70
Pears,apples	50
Stonefruits	50
Tomatoes	40

POST-HARVEST HANDLING STRATEGIES

Curing of Roots, Tubers and Bulb Crops

Curing is required to increase the shelf life of roots and tubers when they will be kept for an extended period of time. In order to heal the skins damaged during harvesting, the roots and tubers must be exposed to high temperatures and high relative humidity for extended periods of time during the curing process. A new, protective layer of cells is created by this process. Although the process of healing initially costs money, it is ultimately beneficial. The conditions for curing roots and tubers are presented in Table 5.

Table 5. Conditions for curing roots and tubers

Commodity	Temperature (°C)	Relative Humidity (%)	Storage time (days)
Potato	15-20	90-95	5-10
Sweet potato	30-32	85-90	4.7
Yams	32-40	90-100	1-4
Cassava	30-40	90-95	2-5

Curing can be done in the field or in facilities that have been prepared for that purpose. The field can be used to cure goods like yams by placing them in a somewhat shaded area. Insulate the pile with cut grass or straw and cover it with canvas, burlap, or woven grass matting. It will be warm enough under this covering to reach high temperatures and high relative humidity. Up to four days can pass with the stack in this condition.

Garlic and onions can be dried in the field in windrows or after being placed inside sizable bags made of fibre or net. In order to create the heat required for high temperatures and high relative humidity, modern curing systems have been developed in housing that is heated and cooled with fans and heaters, as seen below:

The heat is redirected via fans to the lower area of the room where the produce is kept. Bulk bins are stacked with a gap of 10 to 15 cm between rows to allow adequate air passage. Onions can be preserved using the arrangement in Figure 2, however an exhaust opening close to the ceiling is required to allow for air recirculation. The onion bulbs should not become very dry; this should be avoided. Large tarpaulins or plastic sheets must be built into a temporary tent to cure the onions and prevent significant loss when there are harsh field circumstances, such as heavy rain or flooded terrain. The centre of the produce-filled bins has a hollow section into which heated air is forced. The heated air is circulated through the onions while they cure using a number of fans.

The heat is redirected via fans to the lower area of the room where the produce is kept. To ensure enough airflow, bulk bins are stacked with a space of 10 to 15 cm between rows. Onions can be preserved using the arrangement in Figure 2, however an exhaust opening close to the ceiling is required to allow for air recirculation. The onion bulbs should not become very dry; this should be avoided. Large tarpaulins or plastic sheets must be built into a temporary tent to cure the onions and prevent significant loss when there are harsh field circumstances, such as heavy rain or flooded terrain. The centre of the produce-filled bins has a hollow section into which heated air is forced. The heated air is circulated through the onions while they cure using a number of fans.

Activities before packaging

Preliminary procedures are used on fruits and vegetables to enhance appearance and preserve quality. These preliminary procedures include washing, sanitizing, waxing, and coloring.

Cleaning: The majority of product is treated chemically in various ways, such as field spraying with insecticides and herbicides. Even in minute doses, the majority of these compounds are hazardous to people. Therefore, before packing, produce must be free of all chemical residues. The fruit or vegetable is taken to the washing machine and rotated over rotary brushes before being exposed to the cleaning process on all sides.

Fruit is transferred from the washing machine to a set of rotating sponge rollers (similar to the rotary brushes). As the fruit is spun and moved through the sponger, the rotational sponges remove the majority of the water on the fruit.

Disinfection: To prevent the spread of infections among subsequent batches of product, disinfectant agents are applied to the soaking tank after fruits and vegetables have been washed. A typical citrus fruit solution in a soaking tank contains a combination of different chemicals at precise concentrations, pH values, and temperatures, as well as detergents and water softeners. Citrus disinfectants like sodium ortho-phenyl-phenate (SOPP) are efficient, but tank conditions must be carefully controlled. The concentration must be kept between 0.05 and 0.15 percent, the pH must be 11.8 and the temperature must be between 43 and 48 degrees Celsius. It is advised to soak for three to five minutes. If these instructions are not followed, the produce may suffer greatly because the solution won't work if the temperature or concentrations are too low. For several vegetables, low chlorine solution concentrations are also employed as a disinfectant. This solution has the benefit of not leaving any chemical residue on the product.

Synthetic waxing: Produce is coated with artificial wax to replace the natural wax lost during fruit or vegetable washing. The product gains a vibrant gloss as a result. The purpose of producing vegetables artificially waxed is outlined below:

» Provides a protective coating over entire surface.

» Seals small cracks and dents in the rind or skin.

» Seals off stem scars or base of petiole.

» Reduces moisture loss.

» Permits natural respiration.

» Extends shelf life.

» Enhances sales appeal.

Usage of a brand name

Some sellers mark each fruit with a brand name or emblem using ink or stickers. In some nations (such as Japan), ink is not permitted, but stickers are. Pressure sensitive paper stickers can be applied and dispensed by automatic equipment. The benefit of stickers is that they are simple to remove.

Packaging

Modern packaging must comply with the following requirements:

a. The package must be strong enough to protect the contents during handling, shipping, and stacking.

b. The package must adhere to weight, dimension, and shape specifications for handling and marketing.

c. The container should permit quick cooling of the items inside. Additionally, the permeability of plastic films to breathing gases can be significant.

d. High humidity levels or moisture content (when wet) should not significantly impair the package's mechanical strength.

In some marketing scenarios, the safety of the packaging or its simplicity to open and close may be crucial.

a. The container must be transparent or light-proof.

b. The packaging ought to be suitable for shop displays.

c. The packaging should be made to be easily disposed of, reused, or recycled.

d. The price of the package should be as low as possible given its worth and the level of required contents protection.

Classification of packaging

Packages can be classified as follows:

» Flexible sacks; made of plastic jute, such as bags (small sacks) and nets (made of open mesh)

» Wooden crates

» Cartons(fiber board boxes) Plastic crates

» Pallet boxes and shipping containers

» Baskets made of woven strips of leaves, bamboo, plastic etc.

Uses for above packages

Only hard produce, such as coconuts and root crops, should be used with nets (potatoes, onions, yams).Wire-bound wooden crates for citrus fruits and potatoes or wooden field crates for softer produce like tomatoes are the two most common types of wooden crates. Wooden boxes typically have sufficient ventilation and are weather-resistant and more effective for large fruits like watermelons and other melons. The produce may be harmed by rough surfaces and splinters, and when painted, they may retain offensive odors. Additionally, untreated wood is quickly infected by mould. To carry tomatoes, cucumbers, and ginger, fiberboard cartons are employed. They can make product more appealing to consumers because they are simple to handle, low weight, come in various sizes, and come in a range of colours. They have various drawbacks, including the influence of excessive humidity, which can weaken the box; also, they are not waterproof, necessitating the drying of wet materials before packaging. Even though multiple thickness trays are routinely utilised, these boxes frequently have less strength than timber or plastic crates. They may be flat packaged, with grasp handles and ventilation holes, and are a cheap, appealing option that is very well-liked. It is important to take precautions to ensure that the holes on the surface (top and sides) of the box allow for enough production ventilation and prevent heat generation, which can hasten product deterioration.

Although more expensive than wooden or cardboard crates, plastic crates are more durable. Due to their smooth surface and strength, they are easy to clean and provide products with protection. The cost of transportation is decreased by the multiple uses of plastic crates. They are resistant to inclement weather and come in a variety of sizes and colours. However, because of their rough surfaces, plastic boxes can damage some soft produce; therefore, utilizing such crates is advised.

For moving fruit from the field to the packinghouse or for handling produce inside the packinghouse, pallet boxes are particularly effective. Pallet boxes have conventional heights and a standard floor dimensions (1200 x 1000 mm) depending on the commodity. The pallet box has the advantages of requiring less labour and money to load, fill, and unload; requiring less storage space; and speeding up mechanical harvest. The main drawback is that most pallet boxes' return volumes are equal to their full loads. The forklift truck, trailer, and handling equipment that empty the boxes also require a larger expenditure. Due to the substantial initial capital investment, they are not within the price range of small producers.

Minimum temperature advised to lengthen storage

Because every fruit and vegetable reacts differently to cold temperatures, there is no one storage method that works best for all of them. Along with the necessary amount of storage time, consideration must be given to problems like mould growth and chilling injuries. Depending on how perishable they are, fruits and vegetables can be stored at a temperature between -1 and 13 °C. Extremely perishable fruits, such as apricots, berries, cherries, figs, and watermelons, can be kept at a temperature of −1 to −4 °C for 1 to 5 weeks. Less perishable fruits, like mandarin, nectarine, ripe or green pineapple, can be kept at a temperature of −5 to −9 °C for 2 to 5 weeks. Bananas can be kept at −10 Asparagus, beans, broccoli, Brussels sprouts, and other highly perishable vegetables can be kept for up to 4 weeks at -1-4°C for 1-4 weeks and for 2-4 weeks at 5-9°C for cauliflower. Non-perishable vegetables like carrots, onions, potatoes, and parsnips can be stored at 5-9°C for 12-28 weeks whereas green tomatoes, which are less perishable, can be stored at 10°C for 3-6 weeks. Sweet potatoes can also be kept at 10°C for 16 to 24 weeks. Produce's storage life is highly variable and correlated with its respiration rate; there is an inverse relationship between the two, with lower respiration rates often resulting in longer storage times.

A non-perishable fruit like an apple has a respiration rate of 25 mL CO_2. $kg^{-1}h^{-1}$ at 15°C contrasted to a very perishable fruit like a ripe banana, which has a respiration rate of 200 mL CO_2.$kg^{-1}h^{-1}$ at 15°C.

Elevated temperatures: Fruits and vegetables' storage or marketable life is shortened when they are exposed to high temperatures after harvest. This is because, being living things, they often have higher metabolic rates at higher temperatures. High temperatures are useful for treating certain fruits' illnesses and pests as well as drying out bulbous crops and curing root crops. To start or enhance ripening or skin colour, many fruits are subjected to high temperatures in conjunction with ethylene (or another compatible gas).

Storage

Most fresh vegetables can have their commercial life extended by being stored quickly in a setting that preserves product quality. Facilities where temperature, airflow, relative humidity, and occasionally atmospheric composition can be manipulated can provide the appropriate environment. Storage spaces can be divided into two categories: those that need refrigeration and those that don't. In situ, sand, coir, pits, clamps, windbreaks, cellars, barns, evaporative cooling, and night ventilation are examples of storage spaces and techniques that don't require refrigeration.

In situ: Fruits and vegetables are harvested later and kept in storage until they are needed. It can be applied in some circumstances to root crops like cassava, but it prevents the planting of other crops on the area where the crop was previously produced. The crop may be vulnerable to freezing and chilling damage in cooler climes.

Sand or coir: This method of storage, which entails burying the product with sand, is utilized in nations like India to keep potatoes for extended periods of time.

At the boundaries of the field where the crop was produced, pits or trenches are excavated. Pits are typically positioned at the highest point in the field, particularly in areas with heavy rainfall. The crop to be stored is placed inside a hole or trench that has been lined with straw or another organic material, which is then covered with a layer of soil before another layer of organic material is added. Straw is used to make holes at the top for air circulation because inadequate ventilation can lead to crop rot issues.

Clamps: In certain nations, like Great Britain, this has been the customary way to store potatoes. A typical design makes advantage of a plot of land adjacent to the field. The clamp's breadth ranges from 1 to 2.5 meters. The proportions are drawn out, and a long, conical pile of potatoes is placed on the ground. Before planting potatoes, straw is occasionally added to the soil. The heap's angle of repose, which is approximately one third the breadth of the cluster, determines its central height. Straw is bent over the ridge at the top so that rain will often run off the building. When compressed straw thickness should range from 15 to 25 cm. The clamp is covered in soil to a depth of 15-20 cm after two weeks, though this may vary according on the climate.

Wooden pegs spaced one meter apart in two parallel rows are inserted into the ground to create windbreaks. A wooden platform, frequently made out of wooden boxes, is erected between the stakes at a height of 30 cm. between the stakes and across both ends of the windbreak, chicken wire is fastened. In Britain, onions are stored using this technique.

Cellars: These partially buried or underground chambers are frequently seen below a house. Good insulation at this point allows for cooling in warm environments and protection from dangerously low temperatures in cold ones. Apples, cabbages, onions, and potatoes have all been traditionally stored in British homes during the winter in cellars.

Barns: A barn is a type of farm building used for housing, processing, and storing farm animals, produce, and equipment. Even if there is no exact scale or measurement for the kind or size of the building, the term "barn" is typically only used to refer to the biggest or most significant building on a given farm. Sheds or outbuildings are common names for smaller or less significant agricultural structures that are typically used to hold more compact tools or activities.

Aeration by evaporation: Energy is needed for water to evaporate from the liquid phase into the vapour phase. By initially passing the air delivered into the storage room via a pad of water, this concept can be employed to chill stores. The air's initial humidity level and the effectiveness of the evaporating surface determine how much cooling occurs. A significant drop in temperature can be attained by humidifying the low-humidity ambient air to a relative humidity (RH) of about 100 percent. This may offer cool, humid conditions for storage.

Ventilation at night: The difference in temperature between day and night can be utilized to keep shops cool in hot climes. When the harvest is placed inside the storage area, it must be well insulated. The store room has a built-in fan that turns on when the temperature outside drops below the temperature inside during the evening. When the temperatures are balanced, the fan turns off. A differential thermostat, which continuously compares the internal storage temperature with the outside air temperature, controls the fan. Onions are stored in bulk using this technique.

Gastight compartments with insulated walls, ceilings, and floors make up controlled atmospheres. They are becoming more and more popular for larger-scale fruit storage. Different combinations of O_2, CO_2, and N_2 are needed, depending on the species and variety. Fruits with lengthy storage lifetimes are stored in low content02 atmospheres (0.8 to 1.5 percent), also known as ULO (Ultra-Low Oxygen) atmospheres (e.g., apples).

Pest management and decay

Before storage or marketing, crops might be submerged in hot water to prevent illness. This method can be used to successfully manage the common fruit disease anthracnose, which is brought on by the fungus *Colletotrychum spp.* controlling disease in fruits after harvesting is frequently successful when the right amounts of fungicides are combined with hot water. Conditions of storage can also contribute to fruit and vegetable degeneration. When fruits and vegetables are refrigerated, too low temperatures might harm them. High temperatures can increase bacterial infections and soften tissues. Fresh fruits and vegetables are primarily harmed by microbes through the partial or complete physical loss of edible substance.

Chapter - 4

Packaging of Horticultural Products

Poor handling and inadequate packaging are to blame for some of the fresh food post-harvest losses in less developed nations. Other objectives, including as market penetration and competitiveness, are also served by more advanced marketing systems packages. Reduced bruising and crushing as well as improved marketing of food can be achieved by proper packing. Moisture loss and rotting organisms can be prevented by proper packaging, as well as pilferage and hygienic conditions.

When considering the introduction of new packaging, all components of the package must be considered. Cost of packing material, labour costs, acceptance by traders and consumers, and variations in product condition are just a few of these factors. The ultimate purpose of packaging is to improve the quality and marketability of the product by making it easier to handle and transport.

Packaging has four primary functions: it contains, protects, communicates, and markets the thing it contains.

To contain produce

As an efficient handling unit, easy to be handled by one person. As a marketable unit e.g. units with the same content and weight.

a. To protect produce against:

>   Rough handling during loading, unloading and transport-rigid crate.

>   Pressure during stacking.

» Moisture or water loss with consequent weight and appearance loss.

» Heat: air flow through crate or box via ventilation holes.

» Fumigation possible through ventilation holes.

b. To communicate:

» Identification: a label with country of origin, volume, type or variety of product etc. printed on it.

» Marketing and advertising: recognizable trade name and trademark.

c. To market the product:

» The look of fruits and vegetables will improve as a result of proper packaging.

» The marketing of a product will move more quickly and efficiently if it is sold in standard units (weight, count).

» Packaging can be stacked on top of each other in order to save money on transportation and handling. Marketing expenses can be minimized by making better use of available space and preventing waste.

» The use of labels and slots makes inspections easier.

Packaging Materials

Requirements and functions of food containers:

The following are among the more important general requirements and functions of food packaging materials/containers:

i. They must be non-toxic and compatible with the specific foods.

ii. Sanitary protection.

iii. Moisture and fat protection.

iv. Gas and odour protection.

v. Light protection.

vi. Resistance to impact.

vii. Transparency.

viii. Tamper proofness.

ix. Ease of opening.

x.   Pouring features.

xi. Reseal features.

xii. Ease of disposal.

xiii. Size, shape, weight limitations.

xiv. Appearance, printability.

xv. Low cost;

xvi. Special features.

Primary and Secondary containers

Secondary containers have been referred to as primary containers. Some items, such as nuts, oranges, and eggs, have primary containers given by nature. In order to keep the units together and provide general protection, we just require a secondary outer box, wrap, or drum when packaging them.

Primary containers such as plastic liners are often used to fill other items including milk, dried eggs, and fruit concentrates, which are then packaged in protective cartons or drums. Primary container requirements are considerably reduced by the use of a secondary container such as a carton or drum.

Secondary containers, except in specific cases, are not designed to be extremely impermeable to water vapour and other gases, especially at sealing points, relying on the primary container for this function.

Primary containers, on the other hand, are those that have direct contact with the food, thus they will be the focus of our attention.

Hermetic closure: Hermetic and non-hermetic closures are two of the most important packaging criteria. All gas and vapours are completely impervious to the hermetic container, including its seams. If the container is intact, it will be resistant to bacteria, yeasts, moulds, and dirt from dust or any other source because these agents are far larger than gas or water vapour molecules. However in many cases a container that stops microorganisms from entering will not be hermetic. In addition to protecting the product from moisture gain or loss and oxygen pickup from the atmosphere, a hermetic container is necessary for rigorous vacuum and pressure packaging. Metal cans and glass bottles are the most popular hermetic containers; however improper closures can render them non-hermetic. Due to one or more of the following, flexible packages are rarely hermetic.

Although gas and water vapour transfer rates may be exceptionally slow, the seals are generally good but imperfect; and third, even where film materials may be gas and water-vapour-tight, such as certain gauges of aluminium foil flexing of packages and pouches leads to minute pinholes and crease holes, even if the films are completely gas and water vapour impermeable. Hermetic aluminum containers can be easily produced without side seams or bottom end seams, making them ideal for products that need to be completely sealed. In this case, the only seam that has to be hermetically sealed is the top-end double seam, which may be done using standard can sealing equipment. The tightness of the lid is all that's needed to make a glass container hermetic. Plastic or cork inner rings will be used in the construction of lids. Many glass containers are vacuum packed, and the pressure difference between the inside and outside of the container will increase the tightness of the seal. A gas-tight hermetic seal can be achieved by crimping the covers, as in the case of pop bottle caps, which function against positive internal pressure. Bottles, on the other hand, fail to remain hermetic far more frequently than do cans.

PROTECTION OF FOOD BY PACKAGING MATERIALS

Films and Foils Plastics: For example, film and foil have varied values for moisture and gas permeability, strength and flexibility as well as inflammability, as well as resistance to insect penetration depending on the thickness of the film. Table 1 provides an overview of the most popular types of food packaging materials.

Table 1. Properties of packaging films

Material	Properties
Paper	Strength, rigidity, opacity, printability.
Aluminum foil	Negligible permeability to water-vapour, gases and odours, grease proof, opacity and brilliant appearance, dimensional stability, dead folding characteristics.

Cellulose film (coated)	Strength, attractive appearance, low permeability to water vapour (depending on the type of coating used), gases, odours and greases, printability.
Polythene	Durability, heat-sealability, low permeability to water-vapour, good chemical resistance, good low-temperature performance.
Rubber hydrochloride	Heat-seal ability, low permeability to water vapour, gases, odours and greases, chemical resistance.
Cellulose acetate Vinylidenechloride	Strength, rigidity, glossy appearance, printability, dimensional stability. Low permeability to water vapour, gases, copolymer odours and greases, chemical resistance, heat-seal ability.
Polyvinylchloride	Resistance to chemicals, oils and greases, heat-seal ability.
Polyethylene terephtalate	Strength, durability, dimensional stability, low permeability to gases, odours and greases.

When it comes to the construction of containers, these films are typically used. As non-rigid containers, their primary purpose is to keep product contained and shield it from the harmful effects of air or water vapour. When it comes to protecting against mechanical damage, thin films have a limited capacity. Variables such as the kind of polymer, the degree of polymerization, and the molecular weight of polymers, as well as the orientation of the polymers in space, are all factors that might affect the properties of these materials.

Copolymers, a relatively new class of plastics, demonstrate the versatility of combining the building blocks that go into plastics. An example of a copolymer is a resin that may be used to make films and other forms of material. Because of the wide number of potential combinations, copolymers are an essential plastic class for food packaging applications.

Plastic Sheets: Dried fruit leathers, in particular, can be packaged in cellophane paper. There are numerous applications for polyethylene sheets. They're transparent, stretchable, and impervious to water vapour and low temperatures. These sheets have the added benefit of being easily heat sealed, which is a significant perk. Sheets and bags are the most common types of use. It's a good first layer of defence for dehydrated goods. Polyethylene will need to be combined with additional materials in order to provide adequate protection against flavour and gas loss.

Receptacles and Packaging in Plastic Materials

In this class there are two categories:

1. Heat-treatable receptacles include cartons and bottles. Using the same basic materials as those mentioned under "plastic sheets," sterilizable bags can be produced for use up to 120 °C and cellophane for usage up to 100° C. Packing and pasteurization of sauerkraut could benefit from the use of polyethylene bags.

2. Bags and boxes of fruit and vegetables that haven't been heated during processing are also included in this category. A major usage of plastic bags is to package dried or dehydrated fruits and vegetables. Polyethylene or cellophane bags are the most common types of plastic packaging.

In some cases, once the finished product is already within the pack and the air has been evacuated, special packaging can be used to make direct contact with the finished product.

Laminates: In order to combine the best properties of different flexible materials like paper, plastic, and thin metal foils, multi-layers or laminates of these materials are used. These materials combine the best features of each other to provide the best water vapour transmission, oxygen permeability, light transmission and burst strength.

Custom-made laminates with up to eight layers are typical in the commercial laminate industry. Bonding with a wet adhesive, dry bonding with a thermoplastic adhesive or hot melt laminating when one or both layers have the properties of thermoplastics can all be used to create laminations of different materials. It is possible to construct more complicated laminates using these structured plastic films by bonding them to other materials, such as papers or metal foils.

Glass containers: Glass is chemically inert when it comes to food packing; however metal closures present the normal difficulties of corrosion and reactivity. The main drawback of glass is that it is susceptible to breaking, which can be caused by internal pressure, impact, or thermal shock, all of which can be considerably minimized by good container matching and intelligent handling. Consideration here should be taken to consult with the manufacturer. In general, the less likely it is to crack due to internal pressure, the heavier the jar or bottle must be. Thermal shock and impact fracture are both more likely in the heavier jar. The heavier jar is more susceptible to thermal shock fracture because of the wider temperature fluctuations that produce uneven stress on the thicker glass' exterior and inner surfaces. The thicker wall of the heavier jar makes it more susceptible to fracture in an impact.

Each of these types of fracture can be significantly reduced by a variety of coatings. Using specific waxes and silicones, these coatings lubricate the glass. In high-speed filling lines, bottles and jars are less likely to break because they slam into each other instead of glancing off of each other.

Many of the small scratches that develop in regular usage after annealing ovens are protected by surface coating, which also improves the high gloss appearance of glass containers and is reported to reduce the noise from glass to glass contact at filling lines.

Avoiding temperature disparities between the inside and outside of glass containers is a recommended practice when it comes to thermal shock. Some manufacturers recommend that an inside-outside temperature difference of no more than 44° C (80° F) not be exceeded. For a hot fill, bottles need to be warmed slowly and partially cooled before being placed in the refrigerator.

Classification

Glass used for receptacles in fruit and vegetable processing is a carefully controlled mixture of sand, soda ash, limestone and other materials made molten by heating to about 1500°C(2800°F). Main classes of glass receptacles are:

a. Jars which are resistant to heat treatments.

b. Jars, glasses, etc. for products not submitted to heat treatment (marmalades, acidified vegetables etc.).

c. Glass bottles for pasteurized products (tomato juice, fruit juices etc.) or not pasteurized (syrups).

d. Receptacles with higher capacity (flasks etc.).

Jars for sterilized or Pasteurized canned products

Taking into account both the advantages and limitations of these receptacles, metal cans may be replaced by these containers. There are a number of advantages to using these containers: they are non-reactive, transparent, and versatile, and they can be made from a variety of low-cost raw materials. Weight, fragility, and reduced heat conductivity are all drawbacks of this material. It is also less resistant to thermal shocks than a metal can with the same capacity. This category of receptacles necessitates the use of metallic (or glass) tops and particular materials for tightness in order to ensure hermetically after pasteurization / sterilization and cooling. A receptacle's closure method determines what ever type of receptacle it is.

a. Glass jars with mechanical closure.

b. Glass jars with pneumatic closure.

Jars for products without heat treatment

However, glass jars with non-hermetic closures such as those made of metal, glass, or hard plastic lids can also be utilized for these items. Some items (such marmalades and jams) can be filled hot and sterile in containers by using jars with pneumatic closures. In this situation, the receptacles are protected against negative air movement by pneumatic closures.

Glass bottle: These receptacles are widely used both for

a. Finished products which need pasteurization (e.g. tomato juice, Knit juices, etc.).

b. Those which are preserved as such (e.g. fruit syrups).

Glass bottles in category:

a. Are closed hermetically with metallic caps, provided with special materials for tightness.

b. Various corks and aluminum caps with tightness materials may be used.

Glass receptacles with high capacity

Glass flasks with capacities of three and ten liters are available in this category and can be hermetically sealed using the SKO caps system, making them resistant to product pasteurization (e.g. tomato juice).

In order to preserve fruit juices using a heated procedure, glass demijohns with typical capacities of 25 and 50 liters can be utilized as larger containers. Rubber hoods are used to seal the unit.

Paper Packaging

In order to strengthen their protective capabilities, few paper products are used as primary containers without being treated, coated, or laminated. Bleaching and coating or impregnating wood pulp and recycled waste paper with waxes, resins, lacquers, polymers, and aluminums laminations would increase water vapour and gas impermeability, flexibility, tear resistance, burst strength, wet strength, grease resistance and seal ability. *Paper sheets*: For bags and wraps, kraft paper is the brown unbleached heavy-duty paper; it is rarely used as a primary container. water and oil resistance, as well as significant wet strength are all attributed to parchment paper, which is made by treating paper pulp with acid. Wood pulp fibers in glassine-type papers are long and thick, giving them a higher level of tensile strength.

Tin can or tinplate: A tinplate container is referred to as a "tin can." Rigid and impermeable, tinplate is made from a thin sheet of low carbon steel that has been covered on both sides with a very thin layer of tin. Hot-dipped tinplate or an electrodeposited tin layer can be used to make it (electrolytic tinplate). Tinplate with a thicker covering of tin on one side than the other can be created using the second method (differentially coated).

Tin is not completely resistant to corrosion but its rate of reaction with many food materials is considerably slower than that of steel. The effectiveness of a tin coating depends on:

   a.   Its thickness which may vary from about 0.5 to 2.0μm.

   b.   The uniformity of this thickness.

   c.   The method of applying the tin which today primarily involve select politic plating.

   d.   The composition of the underlying steel base plate.

   e.   The type of food and other factors.

Some canned vegetables, such as tomato products, owe their distinctive flavours to a small amount of dissolved tin; otherwise, these products would taste quite different. On the other hand, lacquer may be applied to tin if it has an adverse reaction to a specific cuisine. In order to establish the thickness of tinplate sheets, it is necessary to weigh an area of a certain size and calculate the average thickness. There are many ways to protect the metal surface of cans from corrosion, such as lacquering them after they are made. They can also be printed by lithography on containers made from tinplate sheets to provide necessary instructions or information.

Packaging and Post-Harvest Losses

In the period between harvesting and consumption, post-harvest losses occur. Throughout this manual, the term "losses" refers to any and all losses suffered by the farmer, trader, or consumer (e.g. weight loss, quality loss, financial loss, loss of goodwill, loss of marketing opportunities, loss of nutritional value, etc.). In the Eastern Caribbean, there is a complicated system for distributing fresh vegetables. It takes a long time for the food to get from the farm to the customer, and many small dealers are involved in the process. Due to these and other issues, as well as the high perish ability of fresh produce; there is a lot of food waste and wasteful losses. Post-harvest losses are often categorized according to the principal cause of the losses. Following harvest losses are often the consequence of various reasons, as well as numerous malpractices in the supply chain.

Mechanical Injuries

Losses caused by mechanical injuries include cuts, bruises, abrasions and punctures and can be categorized into four major types of injuries.

Impact injuries resulting from:

Dropping the product on to a hard surface; dropping the product into the back of a car; excessive drops during loading and unloading; suddenly stopping or accelerating a vehicle.

Loading and unloading is mostly done by hand, where by crates are thrown into the pickup, on board and into the hold. Proper, rigid packages possibly with cushioning of each item within the package can reduce these impact losses.

Vibration or abrasion injuries result when produce is able to move within a container because of:

Fields in isolated hilly areas cannot always be reached by vehicles, so that produce has to be carried from the field to the road or track, which is of Leninabad condition. Trucks and pick-ups used on these roads are also often in a bad condition. Tight filling of the crates can decrease the vibration of produce within the boxes and consequently reduce the injury. If the box is not completely filled with produce, it is suggested to use for instance shredded paper to tight-fill the box. Over-tight filling can lead to compression injury.

Compression injuries are caused by improper packing and inadequate package performance resulting from:

Over packing of crate sand boxes too high stacking of crates weak packaging. A high stack of (weak) crates or baskets leads to bulging and consequently to compression of the produce inside. Rain or sea water can weaken the strength of carton boxes resulting in more compression injuries on the produce.

Puncturing injuries resulting from:

» Nails or splinters from the crate or box fingers or nails of a person.

» Other crates, fork-lifts etc. hard and sharp stalks of fruit.

Baskets and old wooden crates and some of the plastic crates often have sharp edges which can easily damage the produce. Rigid crates with proper grips can reduce the incidence of puncturing.

Physical and Environmental Factors

Excessive heat, cold, gases, or humidity can cause a variety of physical and environmental losses, including spoilage and decay. Ventilation and heat exchange are necessary to maintain the optimum temperature level, limit air and gas exchange (oxygen, carbon dioxide and ethylene) and minimize water loss in packing.

To provide for proper ventilation, holes in cartons should occupy at least 5% of the total box surface. To avoid water logging and protect product from ethylene and odour absorption, consumer packaging maintains low oxygen and high carbon dioxide levels.

Biological and microbiological losses: Insects, birds, rodents, germs, and other microbes cause or consume biological and microbiological losses. This type of loss can be reduced by proper packaging, stacking, and storage.

Bio-chemical and physiological losses: Toxic and biochemical losses include unwanted chemical reactions and contamination with dangerous chemicals such as insecticides.

If the wood for the wooden boxes is treated poorly, the quality of the produce may suffer. The sprouting of tubers is one example of a loss caused by a physiological reaction.

Indirect Causes of Post-harvest Losses

External variables account for post-harvest losses that are referred to as indirect or secondary. Increased marketability and reduced losses can be achieved through packaging. Post-harvest losses can be attributed to:

a. Demand from consumers. a. Packaging design could be a part of marketing efforts to promote regional foods.

b. Ineffective marketing systems. Since each individual in the marketing chain is accountable for a step in the marketing chain, delays in the selling of produce might occur, resulting in greater losses. To speed up the distribution of food, packaging that is consistent and standardized can help.

c. Facilities. Access to storage, cold rooms, drying, and curing rooms is restricted. These facilities will benefit from more efficient handling and use as a result of better packaging. The price of the produce will be lower as a result of more efficient utilization of these facilities. Fresh food transportation in the Eastern Caribbean is insufficient and expensive. Produce from the farm is transported to the port in small pickup trucks and wooden schooners are utilized for interisland transportation. In the hold, produce can be stacked as high as three meters without any protection from the elements. If the deck cargo isn't protected from the elements, it's suffocated by the lack of ventilation and lack of airflow. Standardized packing cannot be implemented as long as small trucks and wooden schooners are employed for transportation.

d. Policies (e.g., diversification of agriculture, quality requirements, and pricing policy) must be altered. Large imports or Government measures like the ban Barbados placed on mango imports from Dominica and St. Lucia owing to mango seed weevil pest are examples of supply changes.

e. Regulations imposed by governments can compel businesses to utilize specialized packaging, such as the carton boxes necessary on French-speaking is lands.

f. A lack of training and understanding of the marketing system's participants.

g. An infrastructure that is yet underdeveloped (roads, harbor facilities).

h. Transportation costs. As a general rule, inter-island freight costs are charged per unit, regardless of the type of container. For the sake of saving money on transportation, these containers are constructed to be as large and loaded as feasible. Products are being lost at an alarming rate because of the sheer weight and size of the crates.

Packaging Methods for Horticultural Produce

1. Modified atmosphere storage

Due to their busy life styles, consumers were looking for food products that were convenient. The quality of food is now more important than the quantity because of health concerns. The demand for "natural" and "fresh" products free of "dangerous" chemicals surged significantly. In many cases, modified atmospheric packaging (MAP) appeared to be the best method of food preservation because it considerably increased the shelf life of the product without degrading its "fresh" or fresh-like qualities.

Every 10°C rise in temperature results in a 3–4 fold increase in respiration rate. As a result, MAP's objective for fruits and vegetables is to lower respiration in order to increase shelf life without compromising quality. Lower temperatures, lower O_2 concentrations, higher CO_2 concentrations, and the combined use of O_2 depletion and CO_2 enhancement in pack atmospheres can all limit respiration. Both passive and active methods can be used to alter the atmosphere inside a box. In the first instance, the packaged product and the permeability of the packing material both have a significant impact on the rate of change and the ultimate gas composition in the package. It is common knowledge that most meals originate from living things and stay alive after being harvested. For instance, when fruits and vegetables breathe, they use the oxygen in the air around them and release carbon dioxide in the process. Similar to this, many items' natural micro flora use oxygen.

In addition to the metabolic and physiological activities that use the available gases, foods are subject to oxidative reactions during storage, which over time lowers the oxygen content. The biggest drawback of passive atmosphere manipulation is how slowly it creates the desired atmosphere. Sometimes, this might lead to unregulated quantities of oxygen, carbon dioxide, or ethylene, which can have a negative impact on the product's quality.

This issue can be resolved by actively altering the atmosphere. Typically, to achieve this, a vacuum is first created, and then the desired gas combination is added to the box. Active atmospheric modification occurs at the start of storage and is almost quick in comparison to the passive method. If the right barrier material is utilized and there are no leaks from pinholes or faulty seals, the atmosphere should essentially remain unchanged.

In MAP, nitrogen (N_2), oxygen (O_2), and carbon dioxide are the three primary gases utilized (CO_2). At sea level, the atmospheric air contains roughly 78.1 percent nitrogen, 20.9 percent oxygen, and 0.03 percent carbon dioxide.

Nitrogen is an inert, flavor less gas that has no antibacterial properties. It is primarily used to replace oxygen and prevent packaging collapse and is not highly soluble in water. Oxygen is typically avoided in the MAP of many items due to its detrimental effects on the preservation of food quality. However, some goods occasionally require its presence in minute amounts. For instance, many fruits and vegetables need a certain minimum oxygen concentration to maintain their fundamental aerobic respiration process.

Last but not least, carbon dioxide is soluble in both water and lipids, and it becomes more soluble at lower temperatures. The product's CO_2 dissolving could cause the package to collapse. Bacteriostatic in nature, carbon dioxide slows down the respiration of several products. All three gases are widespread, easily accessible, safe, cost-effective, and not regarded as chemical additions. To maximize the beneficial benefits and reduce the adverse ones, the ideal level of each gas for each food product must be identified and employed. The gases indicated above may occasionally be combined with other gases. For instance, carbon monoxide (CO) is occasionally injected to reduce microbial growth.

CO's use is constrained because it is hazardous to people. Similar to how it can be used to stop oxidative browning, sulphur dioxide (SO2) can also be used to manage bacterial and mould growth. Argon and ethanol have both been used to inhibit microbial development and increase the flavour and firmness of tomatoes, respectively.

2. Controlled atmospheric storage of fruits and vegetables

The fruit and vegetable industry's introduction to controlled atmosphere (CA) storage was arguably the most successful innovation of the 20th century. Low oxygen (O2) and high carbon dioxide (CO2) concentrations in the storage atmosphere, along with refrigeration, are frequently used for CA storage. It is better suitable for fruits like apples, kiwifruit, and pears that are tolerant of long-term storage since the gas composition inside a food storage room is continuously checked and adjusted to maintain the ideal concentration within extremely small tolerances. The formation and maintenance of CA involves a number of processes, such as the removal of excess CO2, the addition of air to replace the O2 lost through respiration, the removal of C2H4, and occasionally the injection of CO2.

Oxygen (O$_2$) removal: There are a number of ways to remove oxygen from CA storage, including reducing O$_2$ levels by horticulture produce naturally respiring in the early stages of CA, rapidly lowering O$_2$ levels in CA rooms, and rapidly reducing O$_2$ levels by flushing with N2. Flushing with N2 is a popular method for bringing O$_2$ levels down to 3-5 kPa, and breathing can bring them down to the required level.

Carbon Dioxide (CO$_2$) removal: Typically scrubber systems are employed to regulate and eliminate CO$_2$ from the storage atmosphere. An aqueous solution of caustic soda is one of the first chemicals used in the CO$_2$ scrubber for commercial CA storage (NaOH). 2NaOH + CO$_2$ Na$_2$CO$_3$ + H$_2$O are created as the caustic soda solution is cycled in open tubes. Due to the corrosiveness and potential hazards associated with handling and disposal of NaOH and potash, their usage was terminated, which led to the substitution of lime for caustic soda. This technique of CO$_2$ removal relies on the process of absorption whereby hydrated lime transforms into limestone (CaCO$_3$) [Ca(OH)$_2$ + CO$_2$ CaCO$_3$ + H$_2$O]. Although activated charcoal scrubbers are rapidly taking the place of lime scrubber systems, lime scrubbing is still in use. CO$_2$ has also been absorbed by other molecular sieve scrubbers such sodium or aluminums silicate zealots. The last novel technique for eliminating CO$_2$ from storage atmosphere is diffusion units. The device comprises of two independent airflow routes in an airtight container and gas diffusion panels with silicon rubber semi permeable membranes.

Ethylene (C$_2$H$_4$) removal: The elimination of C$_2$H$_4$ is essential because some horticultural produce may undergo physiological abnormalities or induce ripening at very low levels of C$_2$H$_4$ (0.1 ppm). The catalytic oxidation of C$_2$H$_4$ to water and CO$_2$ and the usage of C$_2$H$_4$ absorbing beads are the two commercially available types of C$_2$H$_4$ scrubbers. Some CO$_2$ scrubbers, such as N$_2$ flushing systems that are used to remove CO$_2$, also reduce the concentration of C$_2$H$_4$. Researchers created potassium permanganate-impregnated tiny spherical particles known as C$_2$H$_4$-absorbing bead scrubbers. Additionally, they used a variety of porous inert matrix materials, including alumina, zeolite, and silica gel, in the scrubbers. Through a sealed cartridge that is filled with beads, the air from the storage area is circulated. Since beads cannot be renewed, bead scrubbers should be examined frequently to replace used beads with fresh ones. The comparatively high price of potassium permanganate, therefore, restricts the adoption of this method.

Gas control equipment: To reach the desired gas concentrations, the atmosphere in CA storage needs to be managed. In CA storerooms, O$_2$ and CO$_2$ concentrations can be changed independently of one another. In order to successfully store CA, the environment in the storage space should be kept at a stable level. Scrubbers are operated by a variety of controller devices, including on/off switches, proportional (P), proportional-integrated (PI), or proportional-integral derivative (PID), and PC-based controllers. PID controllers are more sophisticated systems that are more intelligent and frequently deployed in CA operations. Because they may automatically change controller parameters, PID controllers have the potential to operate as self-learning or self-tuning systems. O_2 and CO_2 levels in CA storage rooms can be managed online using PC-based solutions. PC systems typically consist of a computer, a data acquisition system, communication ports, switching devices, and control software, among other components. PC systems can be programmed to compute online parameters to automatically log information for subsequent retrieval and analysis or manage interactions between process factors in addition to controlling storage conditions. To enable the operator to enter the necessary control settings and to monitor the operation, even in a remote mode, a number of commercially available dedicated systems are interfaced with a PC.

3. Active packaging

While maintaining the quality of the packaged food, active packaging modifies the state of the food to increase shelf life or to enhance safety or sensory qualities.

The two types of active packaging techniques—absorbers (also known as scavengers) and releasing systems—can be used to preserve food while also enhancing its quality and safety. Systems that absorb (scavenge) unwanted chemicals like oxygen, carbon dioxide, ethylene, too much water, taints, and other particular substances. Releasing systems actively add or emit substances like carbon dioxide, antioxidants, and preservatives to the packaged food or into the headspace of the packaging. Absorbers and releasers might come in the shape of a sachet, label, or film depending on the physical structure of active packaging systems. The head-space of the package is left open for the placement of sachets. Labels are affixed to the package's lid. Direct contact with food should be avoided because it interferes with system performance and, conversely, could result in migration issues.

4. Intelligent packaging

Temperature, gas composition, humidity, and mechanical damage are primarily relevant to maintaining, controlling, and/or monitoring in the packaging needs for fresh fruits and vegetables. In addition to these technical purposes of preservation and protection, packaging also serves as a means of consumer communication, a tried-and-true instrument for marketing, and ease of use.

Time–Temperature Indicators (TTIs)

The most frequent way TTIs provide a visual indicator in reaction to temperature history throughout distribution and storage is by changing colour or moving the colour front. TTIs are often little self-adhesives affixed to, or small devices built into, the package. The shipping container or each consumer shipment may carry these signs. TTIs can be divided into three categories: complete history (which begin integrating as soon as they are triggered), partial history (which begin integrating when a specific temperature threshold is reached), and critical temperature or abuse indicators (that show a change if a certain temperature was achieved and whose response does not depend on the time the product has been packaged).

Enzymatic reaction, polymerization, melting point, and material diffusion are some of the main mechanisms on which TTIs are founded. In the first instance, an enzymatic reaction occurs after the indicator is turned on, changing the pH and, as a result, the colour. The pace at which the polymer coating darkens on indicators that function through polymerization is temperature-dependent. The polymer's colour is contrasted with a printed colour reference surface on the indicator. These indicators typically require deep low temperature storage before use because they are self-activating.

Fresh-cut salads, which require excellent temperature regulation, are one example of a little processed product that the TTIs in the fruits and vegetables sector have a particularly intriguing application for: An acceleration of the pace of decomposition, a slimy texture, and off-odors and off-flavours are all caused by a temperature that is greater than is ideal.

Breathable materials

Each fruit and vegetable has its own ideal environment, which when combined with a controlled temperature, helps to preserve the quality and freshness of the produce. The respiration rate will increase and the amount of oxygen consumed may exceed the package's capability to allow for oxygen entry if the temperature control fails, though. As a result, the produce will spoil because the oxygen concentration is too low (or, conversely, the carbon dioxide concentration is too high). This problem arises because the respiration rate of produce is more temperature-sensitive than the permeability of the majority of conventional packing films, such as polyethylene.

Breathe Way™ membrane technology, a novel idea for packages with gas permeability that responds to temperature changes to compensate for mild fluctuations happening during transportation. This is a way to establish various optimal oxygen and carbon dioxide levels in the package, maintain these optimal levels within bounds regardless of temperature changes, and provide variable package permeability.

The approach entails applying a highly permeable membrane over a package hole. The membrane is created by applying a proprietary side chain crystals able (SCC) polymer on a porous substrate. In SCC polymers, the side chain crystallizes separately from the main chain. Siloxanes and acrylic polymers are two examples of these polymers. When heated, the polymer transforms from a solid or crystalline state to an amorphous fluid. Since the polymer permeability varies and increases noticeably when moving from below to above this temperature, the temperature of this rapid transition acts as a switch. When the temperature rises by 100C, the temperature switch enables membrane permeability to rise by up to 1.8 times, offsetting the increase in the respiration rate.

Gas and volatiles indicators

We may add leakage indicators, which are based on the detection of oxygen and/ or carbon dioxide, as well as indicators of food spoilage and food quality loss or ageing, usually referred to as freshness indicators. They are based on the detection of volatile metabolites such hydrogen sulfide, carbon dioxide, diacetyl, amines, ammonia, and ethanol that are formed as food deteriorates. The riper sensor label, a ripeness indication created in New Zealand by Jenkins Group in collaboration with Hort Research, is the sole commercial application that has been successfully manufactured to date. When a fruit ripens, it undergoes physiological changes as well as the development of textural and sensory qualities that make it suitable for consumption. With the breakdown of organic acids and the conversion of starch to sugars, there is an increase in sweetness; with the loss of cell turgor, the conversion of pectin's, and the breakdown of cell wall components, there is a decrease in firmness; with the production of aromatic volatile compounds, there is an increase in flavour and with the production and breakdown of pigments, there are changes in colour. Changes in colour in various fruits can be utilized by consumers to determine when the fruit is ripe and "good to eat" (e.g. bananas, avocado, tomatoes).

Kiwifruit and the majority of pear cultivars, for example, don't show any overt visual signs of ripening. This sensor, known as ripe sense, was created especially for pears. The sensor changes colour by sensing the naturally occurring aroma compounds released by the fruit as it ripens, and the container design, created to capture the emitted aroma, shields the fruit from crushing or bruising, allowing retailers to sell ripe, tender, and ready-to-eat fruit without excessive shrinkage. The sensor starts off red and then changes to orange and eventually to yellow. Customers can precisely select fruit that is as ripe as they desire by matching the sensor's colour to their eating preferences.

5. Edible coating and films

To protect fruits and vegetables against moisture, oxygen, and solute movement, edible coatings are thin layers of edible material applied to the product surface in addition to or in place of the natural protective waxy coats. To generate a changed ambiance, they are dipped, sprayed, or brushed directly over the food surface. The substance used to create edible films and coatings must be generally recognized as safe (GRAS), approved by the FDA, and in compliance with the laws governing the relevant food product. Film-forming materials can be used to create edible coatings. A solvent such as water, alcohol, a mixture of water and alcohol, or a combination of other solvents must be used to scatter and dissolve the film materials during manufacture. In this process, plasticizers, antimicrobials, vitamins, minerals, colours or flavours may be added. For the particular polymer, the pH may be changed and/ or the solutions heated to aid in dispersion. To create freestanding films, the film solution is then cast and dried at the proper temperature and relative humidity. There are numerous ways to apply the film solutions on food, including dipping, spraying, brushing, and panning, followed by drying. Polysaccharides, proteins, lipids, and composites can all be found in edible coatings.

Properties of edible coatings

> » The coating must be water-resistant in order to stay intact and completely cover a product after application.

> » There should be no significant carbon dioxide or oxygen depletion. To prevent a switch from aerobic to anaerobic respiration, an area around a commodity must have a minimum of 1-3 percent oxygen.

> » It should improve aesthetics, maintain structural integrity, enhance mechanical handling capabilities, convey active agents (antioxidants, vitamins, etc.), and retain volatile taste compounds. It should also limit water vapour permeability.

> » It must melt at temperatures above 40°C without decomposing.

> » It must never degrade the quality of fresh produce and must not emit offensive odours.

» It should be cost-effective and low viscous.

» It should be able to withstand light pressure and range from being translucent to opaque without being like glass.

Fruits and vegetables that are coated as whole include:

Fruits: Apple, kinnow, grapefruit, passion fruit, avocado, orange, lime, peach, and lemon.

Vegetables: Cucumber, bell pepper, melons and tomato.

Freshly cut foods are very perishable, mostly because the skin—barrier—is nature's removed from their surface area and because of the physical strain involved in peeling, cutting, slicing, shredding, trimming, coring, etc. Commercial coatings for freshly cut fruits and vegetables include:

Fruits: Fresh-cut apple, fresh-cut pear, fresh-cut peach.

Vegetables: Minimally processed carrot, fresh-cut lettuce, fresh cut cabbage, minimally processed onion, fresh-cut potato, fresh-cut tomato slices, fresh-cut muskmelon and cantaloupe.

Nanotechnology in fruit and vegetable packaging

In terms of antibacterial qualities, barrier qualities, and high strength, packaging made of nano particles has numerous advantages over conventional packaging. Titanium dioxide nano particles with photo catalytic properties can oxidize ethylene to produce carbon dioxide and water, avoiding loss from over ripening. Along with its ability to kill germs, silver nano particles also exhibit photo catalytic characteristics. Additionally, ethylene decomposition and contamination prevention have both been accomplished using nano-Ag. Coating Fuji apples with nano materials like nano-silicon oxide (SiOx)/ chitosan has been done in the past. It is feasible to create superior packing in terms of mechanical and thermal barrier qualities by the addition of suitable nano particles. Fruits and vegetable packaging is a constantly evolving industry. Consumer desires for fresher, safer, more nutrient-dense, and convenient produce are a major driving force behind it. Packaging that keeps fruits and vegetables in a fresh state for longer lengths of time during transport is necessary due to an increase in the importation and exportation of fresh commodities. Produce packaging that may prevent, detect, or completely remove fruit and vegetable contamination is becoming more important than ever due to growing concern and awareness about foodborne pathogens and toxins. Therefore, the aforementioned packaging solutions are required to address the numerous problems that the fruit and vegetable sectors face.

Labelling of Packages

Each containers hold be marked with a label with the following information: Country of origin.

» Name and address of exporter or grower. Brand name.

» Description of content (product, variety, size, class, quality grade).

» Gross weight.

» Net weight or count.

» Overall dimensions in metric units. Full name and address of receiver.

The following rules should be maintained with regards to labels:

» There should be at least two labels on each container's short sides.

» The label needs to be positioned so that it is least likely to get dirty or destroyed.

» General information like the brand name, the type of product, and a logo should be printed on each of the container's long sides.

» Additional information, such as FRAGILE, TOP, or specific handling or storage instructions, should be written on the crate's top and at least one of its sides.

» Waterproof ink should be used exclusively.

» Where these extra costs may be justified, colour differences for the various commodities and grades should be employed.

» Remove or tape off any outdated labels.

» The handwritten text on the label should be block style.

   Additional comments, including the date of packaging, legal observations, and labeling for electronic scanning, may be added.

Retail packaging

The following benefits can be obtained by prepackaging or retail packing particular goods in specially created consumer packages:

» The rate of deterioration or degradation is slowed.

» The lessening of product spoilage as a result of consumers picking only what they want.

» Less time will be required for weighing and checking after the consumer makes their choice.

» Advertising through a plentiful supply.

» Greater produce protection.

However, excessive relative humidity can cause roots and tubers to sprout, while low oxygen levels will reduce respiration to the point that produce would suffocate and rot. In order to keep green vegetables from withering, it is generally advised to simply poke a few small holes in the packaging. Because too much relative humidity or too little oxygen causes sprouting, root crops need additional holes or even to be packaged in nets.

Four to six units (citrus, apples, etc.) are combined into one bigger unit on molded trays that are covered in film liners. The vegetables will receive some protection on the edges and some protection from the bottom side of the tray. Only when the larger pieces are correctly placed will the top be adequately covered.

Low-density polypropylene is the material used the most frequently for bagging, regenerated cellulose or cellophane is used to cover the trays, and polyvinyl chloride film is used to cover the trays.

Due to the limited gas and vapour permeability of polyethylene and polypropylene, holes must be drilled into the bags. There are variances in permeability even across various brands of the same material, while other materials have higher permeability's for just one of the gases or vapour. It is advised to buy film material designed for a particular item and to test it out first on a small scale.

Chapter - 5

Hygiene and Sanitation
in Post-harvest Handling

In the last decade, there has been a surge in the number of people worried about food safety when it comes to fresh fruits and vegetables. Berries, tomatoes, leafy greens, and cut fruits and vegetables have been linked to recent outbreaks of food-borne disease. Consumers and wholesalers alike are becoming more concerned about the safety of the food they buy. Growers and post-harvest workers must record their procedures for safeguarding fresh product against contamination. Retailers, such as huge supermarket chains, are increasingly requiring that their suppliers adhere to food safety regulations.

The following are the most common causes and sources of food safety issues during the production and post-harvest handling process.

Physical hazards: Examples of physical hazards which may become imbedded in produce during production handling or storage are such things as: fasteners (staples, nails, screws, bolts).

Chemical hazards: Examples of chemical hazards which may contaminate produce during production handling or storage are such things as: pesticides, fungicides, herbicides, and rodenticide machine lubricants from for klifts or packing line equipment heavy metals (Lead, Mercury, Arsenic) industrial toxins compounds used to clean and sanitize equipment.

Human pathogens: There are four main types of human pathogens associated with fresh produce:

> » Soil associated pathogenic bacteria (*Clostridium botulinum*).

»   Feces associated pathogenic bacteria (*Salmonella spp.*, *Shigella spp.*, *E. coli*).

»   Pathogenic parasites (Cryptosporidium, Cyclospora)

»   Pathogenic viruses (Hepatitis, Enterovirus).

Many of these infections are transmitted via a human (or domestic animal) to food-to-human transmission pathway. Cross contamination, polluted irrigation water, insufficiently composted manure, or contacts with contaminated soil are just a few of the ways in which human infections can be transferred to food. The quality of produce can be evaluated by its outer appearance, including colour, turgidity, and aroma, but this does not apply to food safety It is impossible to tell if a piece of produce is safe to eat simply by looking at it. Growing and post-harvest handling conditions must be carefully controlled to avoid physical risks, hazardous chemicals, and human infections contaminating fresh fruit.

Microbiological Risks

After harvest, products go through a variety of processes. This opens up a slew of other potential sources of contamination in addition to those that occur in the field on their own. Products and packaging with alien materials are fiercely resisted by consumers. Insects, plant waste and animal faces are a few examples of what you might find here. As a result, they are rather easy to identify and remove. The presence of human diseases on produce is a more serious issue. Some of these changes in look, flavour, colour or other exterior features may make them difficult to see or identify. Certain infections have been demonstrated to be able to persist on produce long enough to pose a risk. When it comes to food borne sickness, there have been several reports (Table 1).

Viruses (such as hepatitis A), bacteria (such as Salmonella spp., Escherichia coli, Shigella spp., and others), and parasites can all be transferred on fruits and vegetables and pose a risk to human health (*Giardia spp.*). Mycotoxins and fungi are rarely a concern. Due to the fact that fungus are usually discovered and eradicated before mycotoxins are formed. Bacteria are the most common cause of illnesses linked to eating fruits and vegetables.

Contamination of produce can occur through a variety of causes. The greatest way to ensure the safety of a product is to avoid contamination at all stages. Microorganisms do not thrive in settings that are too hot or too cold, such as in a refrigerator. The "systems approach" is the name given to this method.

All the steps in the process work together to form a single integrated system. All activities and treatments must be documented in order to develop a tracking strategy. A flaw in the system can be discovered and remedial steps taken in this manner. GAP and/or GMP compliance is essential to the successful implementation of this system. With the help of technologies like Hazard Analysis Critical Control Point (HACCP) analysis, food safety hazards can be discovered and mitigated.

This section provides a brief overview of some of the most important aspects that can affect the health and safety of people who eat fruits and vegetables.

Before Harvest

Some infections that affect humans are found in nature. However, human, animal, and wild animal faces are the primary source of produce contamination. Irrigation and washing water are the primary means of entry. It is possible that microorganisms in surface water (lakes, rivers, etc.) are the result of untreated municipal sewage being dumped upstream. Septic tanks may also contaminate underground water by leaking into aquifers through the earth. Underground drip irrigation is the sole suggested method for avoiding contamination of above-ground edible plants with tainted water. It's common for fruits and vegetables to rely extensively on human labour in their cultivation and harvest. Field workers' personal hygiene is another potential source of infection. Bathrooms and other sanitary facilities for the cleanliness of workers are often a significant distance away from manufacturing fields, where contracted teams of migrant Jab our temporarily reside. It is regarded unacceptable to live in such settings or to perform such hygiene habits. Employees must be educated on the need of good hygiene in order to ensure food safety in addition to having portable restrooms on hand. Bacteria and fungi tend to predominate in vegetables with low acidity in tissues, whereas fruits are more likely to be infested by fungi. Strawberries and leafy vegetables in general are more vulnerable to contamination by water, soil, or animals than tree crops, which are more resistant to contamination. Tissue chemical elements such as organic acids, essential oils and pigments have anti-microbial properties and help to prevent the growth of microbes. Harvesting presents a plethora of contamination risks, just like any other handling process. Microorganisms can thrive on the latex and other plant substances exuded by wounds and bruises, which can then be spread via hands, tools, clothing, water or containers. The conditions to which the produce is exposed might aggravate contamination at any stage along the distribution chain. Taking temperature into account is the most crucial consideration.

Market preparation

Concerns raised in the preceding paragraphs about product handling and personal hygiene apply just as much to the process of getting a product ready for sale. However, there are a few more things to consider. People who are ill or have open wounds should not be allowed to come into touch with the product in packing houses or processing plants. When handling the product, employees should wear hairnets and sterile clothing to prevent the spread of infectious diseases. There should be no eating or drinking in the packing area, and workers should wear their work uniforms. Employees should wash their hands every day before they begin work and again after they leave the restroom. Water, on the other hand, is more likely to be a contaminant throughout the manufacturing process. In a packinghouse, water is used for a variety of purposes, including product washing, container cleaning, and hydro-cooling. Personal hygiene and usage as a carrier for waxes, chemicals, and the like are other possible applications.

Water disinfection

Minerals and dissolved gases are among the many water pollutants that need to be addressed. These include suspended materials, bacteria, organic debris, off-colors, and off-odors. To ensure food safety and interaction with food, municipal water is filtered and treated (usually with low chlorine concentrations) to meet the chemical and microbiological criteria. The water from other sources must be filtered and sanitized before it can be consumed. With the usage of municipal water, sanitation is still essential in order to prevent contamination and the transmission of disease to neighboring units. Water can be disinfected in a variety of ways. Chemical, thermal, ultrasonic, and irradiation are all examples of these. There are many ways to get rid of fungi and bacteria after harvest, but chlorine and its derivatives are the most commonly employed. Chlorine is a highly reactive gas with a powerful and pervasive odour. At the post-harvest level, it is utilized mostly in three forms: as pressurized gas from metal cylinders, as calcium hypochlorite (solid) or liquid as sodium hypochlorite, widely known as "bleach" for household whitening and sanitizing. In most cases, chlorine gas is used in large-scale processes, such as municipal water treatment. Concentrations of 65 percent calcium hypochlorite are common, however cold water makes dissolving it more difficult than hot water. As a result, sodium hypochlorite costs more per unit of chlorine concentration than the other two formulations (5 to 15%). It is, nonetheless, useful for small-scale activities because of its simple dosing.

Hypochlorous hypochlorite ion or combinations of both are all forms of chlorine that can be found in aqueous solutions, depending on the pH of the solution. In comparison to hypochlorite, hypochlorous acid has a bactericidal effect 50 to 80 times greater. The pH solution should be in the range of 6.5 to 7.5 for maximum effect on bacteria. The hypochlorous form is particularly volatile and prone to escaping as a gas, resulting in irritation and discomfort for employees below this point. Corrosion is also a problem for the equipment. However, its usefulness as a sanitizer diminishes dramatically above 7.5.'s concentration. Sodium hydroxide can be used to alkaline, while vinegar can be used to acidify a solution. Hypochlorite maintenance kits, either calcium or sodium, will be increased. Parts per million (ppm) is the unit of measurement for active chlorine concentration (ppm). Most bacteria and fungi can be killed by concentrations of active chlorine in the range of 0.2 to 5 ppm. For washing and hydro-cooling, greater concentrations (100-200 ppm) are employed in commercial operations. For example, 80 g active chlorine/dm3 in 400 liters of water is equivalent to 200 parts per million, while 100 and 50 parts per million are used in 800 and 1600 liters, respectively. You can begin daily operations with low concentrations (between 100 and 150 parts per million), and when water becomes contaminated with dirt and plant debris and bacteria increases, you can add more chlorine solution. 3-5 minutes of exposure is sufficient for disinfection purposes. But in addition to pH and contaminants, temperature is also a consideration. Low temperatures have the potential to inhibit activity, thus the reason. It's also vital to consider how well bacteria can grow. As a result of their spores' 10 to 1000-fold greater resistance to death in their dormant stage.

Some countries forbid the use of chlorine in the production of fruits and vegetables. As a result of this, chlorate compounds can be formed when it reacts with organic materials. These have been linked to cancer. As a result, the sanitizer business is looking into other options.

Concentrations of ozone range from 0.5-2 ppm, and it has a severe oxidizing effect. It has been given the go-ahead for use in water treatment facilities. It is, however, difficult to put this into practice. This is due to the lack of currently available technologies for reliably monitoring concentration levels. To make matters more complicated, it only works in a pH range of 6-8 and must be created on-site. Concentrations more than 4 ppm are harmful to humans, and some plant tissues may be damaged as a result. Despite these drawbacks, it appears to be the most promising substitute for chlorine at this time. An ultraviolet light source of 250-275 nm can also be utilized. The pH or temperature of the water has no effect on it. However, because turbidity diminishes water's effectiveness, it must be filtered. The management of water is also crucial. This is due to the fact that multiple washings are necessary. In comparison to a single wash, this is more effective. Good cleaning procedures can be summarized as follows: Once soil and plant debris are removed, go to the next step. Second, use chlorinated water to wash, and then rinse with water that isn't chlorinated. Washing is more efficient when the water is agitated with a brush or a paddle. Washing water can be reused after rinsing; however it must be done in the opposite direction of the product's flow. For pre-cooling, hydro cooling is among the most effective methods. It is, however, one of the most susceptible to microbial contamination of all the options available to us. If there's water in the fruit, this could be the problem. It's crucial to think about forced air as an alternative pre-cooling option because of this.

Plant hygiene

As a result, industrial facilities adhere to a rigid set of hygiene regulations. However when preparing vegetables for the fresh market, hygienic facilities are sometimes overlooked. If low-cost materials have been employed in the packing shed's construction, this is especially true. The packing shed should be constructed to allow for comprehensive cleaning procedures, but other considerations must be made before designing and organization can begin. It's best to keep the delivery area isolated from the reception area. A similar principle applies to the separation of "clean zones," or places where product preparation takes place, from those used to handle receivables from the field. Taking breaks, changing clothing, and taking care of personal hygiene should be easy for employees in a designated place. Even if dust and other pollutants are removed, the facility and equipment, especially those in touch with produce, should be disinfected using liquid sanitizers. Sanitizers based on chlorinated hydrocarbons are the most often used disinfectants. However, the type of water, pH, cost, and the type of equipment all play a role in the final decision. Disinfectants based on iodine (iodophors) are less damaging to metals than chlorine-based disinfectants. However, their efficiency is limited by the pH range in which they operate (2,5-3,5). Stains can also be caused by them. Quaternary ammonium compounds are commonly used for cleaning floors, walls, and aluminium equipment. As a result, they are non-corrosive and do not react with biological materials. It is also pricey and leaves behind residues on the surface, which is not ideal. In the food industry, there are additional sanitizers that can be utilized. There are a wide variety of animals that can disperse microbes in the environment through the excrement they produce. It should be illegal for anyone, including pets, to enter packaging areas. Insect-proof screens should be installed on all doors, windows, and other openings to prevent them from becoming entry points. The use of approved pesticides, traps, and baits is also critical to preventing the spread of insects and rodents. Keeping the surroundings and the facilities clean and tidy is essential if you want to keep insects, rodents, and other animals out of your facility. Waste and trash should be removed on a daily basis.

Storage and transport

Human hygiene contamination of workers and facilities is also applicable here in terms of sanitation risk. The use of new containers and the avoidance of repackaging are two more recommendations. Fruits and vegetables should not be stored or transported in close proximity to other fresh food items.

Keeping products at the appropriate storage temperatures is the most effective technique for preventing the growth and development of human pathogenic organisms. There are three primary groups of microorganisms based on their temperature tolerance. There are three types: psychotropic, which can grow in refrigerators even though the ideal temperature for growth is between 20 and 30 degrees Celsius; mesophilic, which thrives in temperatures between 20 and 40 degrees Celsius; and thermophillic, which needs temperatures above 40 degrees Celsius. C and d aren't directly related to fresh fruits and vegetables, but they may be found in under-processed products. Psychrotrophics can grow on produce that has been stored for a lengthy period of time if refrigeration is not used.

Microbial growth is also influenced by the environment in which a product is stored. There is no need to be concerned about *Clostridium botulinum*, for example, when preparing a product for the fresh market. Tissues with a pH of greater than 4, 6 and low oxygen levels, on the other hand, may form and create poisons. It can be found in canned goods that have been improperly pasteurized, but it can also develop in an environment with a changed atmosphere. This bacterium has been linked to human poisoning.

Sale

Fruits and vegetables can be infected at the point of sale, in storage, and in the kitchen at home. The prior discussions on personal cleanliness and animal avoidance are still relevant here. Large fruits like pumpkins, watermelons, and melons (among others) should not be chopped into sections at retail because of the frequent retail practice of doing so, and refrigeration is required for perishable products.

Food Safety on the Farm

Practices related to these four simple principles can reduce the risk that produce may become contaminated on the farm.

Clean soil

»   Avoid the improper use of manure.

»   Compost manure completely to kill pathogens, and incorporate it into soil at least two weeks prior to planting.

»   Compost manure completely to kill pathogens, and incorporate it into soil at least two weeks prior to planting.

»   Keep domestic and wild animals out of fields to reduce the risk of fecal contamination.

»   Provide portable toilet facilities near the field.

»   Prevent run of for drift from animal operations from entering produce fields. Do not harvest produce within 120 day so for manure application.

Clean water

»   Test surface water that is used for irrigation for fecal pathogens on is basis, especially if water passes close to a sewage treatment or lives to area.

»   Keep lives to a way from the active recharge are for well-water that will be used for irrigation.

»   Keep chemicals away from the active recharge area for well-water that will be used for irrigation.

»   Filter or use settling ponds to improve water quality.

»   Where feasible, use drip irrigation to reduce crop wetting and minimize risk. Use potable water for making up chemical pest management sprays.

Clean surface

»   Tools and field containers must be kept clean. Wash and sanitize these items before each use.

Clean hands

» Workers who harvest produce must wash their hands after using the toilet.

» Provide soap, clean water and single-use to wells in the field and insist that all workers wash their hands before handling produce.

Proper hand-washing is an effective strategy for reducing risk of contamination, but food safety experts have observed that few people wash their hands properly. Cornell's Good Agricultural Practices Program provides the following steps:

» Wet hands with clean, warm water, apply soap and work up lather.

» Rub hands to gather for 20 seconds.

» Rinse under clean, running water.

» Dry hands with a single use towel.

Minimizing Pathogen Contamination during Harvest

A dirty hand or knife blade can easily infect fresh fruits and vegetables during harvesting activities. All members of the harvest crew should have access to and usage of portable field latrines and hand wash stations. Field worker hygiene behaviors, such as washing hands after using the toilet, must be closely monitored and strictly enforced if human pathogen contamination is to be avoided. Harvesting fresh produce should not be assigned to workers who have hepatitis A or who show signs of nausea, vomiting, or diarrhea. After harvest, produce should not be left on bare soils but instead should be stored in hygienic field container. Agricultural harvesting instruments and gloves should be disinfected and sterilized before use. Regular cleaning and sanitization of field containers is necessary to keep them free of contaminants like mud, industrial lubricants or metal fasteners. Reduce the transmission of pathogens by preventing workers from standing on harvesting bins. Because of the ease with which they can be cleaned and sanitized after each usage, plastic field bins and totes are favoured over wooden containers. After each usage, containers should be washed and sanitized to prevent contamination, which could then spread to the next product that is placed in the container. It is nearly impossible to disinfect wooden containers or field totes, since they have a porous surface, and wooden or metal fasteners, such as nails, from wooden containers may accidently be transferred into the produce itself. To avoid the possibility of cross-contamination, cardboard field bins should be visually inspected for cleanliness and lined with a polymeric plastic bag before reuse.

In certain cases, produce may be field packaged in containers that will go to the final destination market, while in other cases it may be brought to a packing shed in bulk containers, baskets, or bags. Any water that comes into contact with the gathered produce must be sanitary and clean.

Minimize Pathogen Contamination during Postharvest Handling

Employee hygiene

Employees of packing sheds geared toward export are frequently seen donning protective clothing such as gloves, hairnets, and clean smocks. To limit the possibility of infection, it's critical to keep workers handling produce clean and hygienic at all times. In order to avoid packinghouse workers from contaminating fresh fruit, ensure that there are adequate restrooms and hand washing stations. There may also be shoe or boot cleaning stations to decrease the quantity of dirt and contamination from the field that reaches the packing shed. When an employee is employed, they should receive training on safe food handling techniques, and that training should be repeated before they begin work each season.

Equipment

Conveyor belts dump tanks, and other food contact surfaces should be cleaned and sanitized on a regular basis using cleaning agents certified for food contact surfaces. As a food contact surface sanitizer, sodium hypochlorite (bleach) at 200 parts per million is a great choice. Once organic things like dirt and plant matter have been removed, disinfectants can be utilized. A bio film is formed when organic materials are caked by steam and cannot be sanitized, hence steam cleaning should be avoided at all costs. Bacteria can be transmitted throughout a packing house facility by aerosolizing bacteria from the steam.

Packaging materials

To prevent harmful substances from leaching out of the package and into the produce, all packaging materials should be made of food contact grade materials. Due to the usage of recycled materials, some packaging materials may contain toxic chemical residues. Containers such as boxes and plastic bags that are no longer in use should be stored in a climate-controlled environment to avoid exposure to insects, rodents, dirt, and other contaminants. As a result of these measures, the integrity and safety of priceless resources is not only preserved but also increased.

Wash and Hydro-cooling water

All water used for washing or hydro-cooling produce must be safe for consumption. Chlorine levels in water should be between 100 and 150 parts per million, with a pH ranging from 6 to 7.5. All produce in the washing or hydro-cooling system will be protected from cross-contamination, but it will not be sterilized. Dump tanks and hydro-coolers should have their water changed on a regular basis.

Refrigerated transport

A refrigerated truck is the finest method of transporting produce. Prior to loading, pre-cool the cars. Extending the shelf life of perishables and reducing the development rate of germs, including human diseases, even while they are being transported to their final markets is possible by keeping them below SQC (41Q F). Most pathogens can develop at temperatures used to transport chilled delicate products. Every time a truck is used for transportation, it should be sanitized and cleaned. Never employ a truck that has previously hauled live animals, animal products, or hazardous chemicals for the transportation of fresh produce.

Sanitizing field containers tools and packs house surface

Prior to each day's harvest, thoroughly clean all crop containers, tools, and pack house surfaces using high pressure washing, rinsing, and sanitizing. Cleansers should only be used after a thorough abrasion washing to remove organic elements such as dirt or plant matter. Chlorine and quaternary ammonium compounds are the most common ingredients in commercial sanitizers (QUATS, QAC, benalkonium chloride, N-alkyl dimethybenzyl ammonium chloride). Quaternary ammonium compound sanitizers cannot be used with chlorine solutions made from chlorine gas, hypochlorite, or chloramines. The surface to be cleaned, the hardness of the water, the application equipment available, the effectiveness under ambient circumstances, and the cost all factor into the sanitizer selection process. Compressed gas, powders, and concentrated liquids all require the utmost caution when being handled.

As a comparison to meat and other foods, fruits and vegetables are more microbiologically safe. Since there is no kill step in the process of battling microbes, many organisms are not destroyed (by cooking), making them harmful if contaminated.

It's hard to tell how serious this threat is. For this reason: when anything turns serious, it is usually publicized. Food borne disease isn't linked to fruits and vegetables because they're considered "good foods". Instead, it's more likely that something else you ate that day is to blame. Evidence suggests, however, that this is becoming an increasingly serious issue. There are two possible explanations for this: First and foremost, agricultural methods are moving in the direction of being more environmentally friendly. The use of organic fertilizers or soil additions based on manure increases the danger of contamination. As a result, a single occurrence of contamination can have a significant impact on the entire supply chain because supermarket distribution centers feed a huge number of stores.

The first step in obtaining high-quality produce with minimal danger is to comprehend the intricacy of microbial contamination and recognize its significance. Microbial contamination is a difficult issue, and excellent agricultural and industry methods are shown here to be aware of this. This section's examples of best practices in agriculture and manufacturing may not be applicable to all fruits and vegetables. If you're looking for specific prevention strategies, then these may be useful. With today's technology, there is no way to completely eliminate this risk. Although it's critical that you learn as much as can about how to minimize it, Preventing microbial contamination of fruits and vegetables is more cost-effective and efficient than dealing with it after the fact. Everyone in the food supply chain, from producers to consumers, must put in the time and effort to ensure food safety. When it comes to preventing contamination, the availability of educated staff and a system that assures no gaps in the quality chain must be considered.

Many different testing procedures exist for the detection of microorganisms, such as total plate counts or aerobic plate count. In terms of analyzing food safety, they provide a rough estimate of microbial contamination; however they are of little use in this regard. A wide range of microorganisms exist naturally on the surfaces of fruits and vegetables and they will colonies a growing medium. This does not, however, imply that they pose a health risk they can be used to monitor the hygiene system or evaluate the effects of various hygienic actions. In order to detect Salmonella and other hazardous germs, certain tests are required, and their absence does not imply that the produce is free of other microbes.

Since the optimal strategy is to minimize the danger and avoid contamination at all costs, this is the best course of action. Tracing is a critical component of any initiative to improve agricultural or industry processes. Contamination issues can be easily identified and located thanks to the technology. Corrective actions can be put in place as soon as the situation warrants it. In the event of an outbreak, quick response is difficult due to the small time span between harvest and consumption of fruits and vegetables. However, despite these drawbacks, maintaining records could aid in reducing the population at risk and should be used in conjunction with the other preventive methods discussed above.

Chapter - 6

Post-harvest Quality
and Safety Management

Fruits, nuts, and vegetables are important sources of vitamins, minerals, dietary fibre, and antioxidants, which are essential for human nutrition. People who consume a wide variety of fruits and vegetables on a regular basis are more likely to avoid cancer, heart disease, and stroke as well as other chronic diseases. Between harvest and consumption, horticultural commodities suffer both quantitative and qualitative losses. For qualitative losses, such as nutritional quality, caloric value and consumer acceptability of fresh produce, it is considerably more difficult to measure than quantitative losses. The marketability of a product and the extent of post-harvest losses are both affected by variances in quality standards, consumer preferences, and purchasing power between countries and cultures. Commodity kinds, producing regions, and seasons all influence post-harvest losses in a variety of ways. Between the production and consumption sites, losses of fresh fruits and vegetables in developed countries are reported to range from 2% for potatoes to 23% for strawberries. As a contrast, developing countries suffer from a wide range of produce losses. Estimated losses at retail, food service, and consumer levels are about 20% in affluent countries and 10% in underdeveloped countries. An estimated one-third of all horticulture crops are never intended for human consumption.

Increasing the amount of food available to the world's rising population, reducing the land required for production, and conserving natural resources can all be achieved through reducing post-harvest losses. Loss prevention methods include:

» Using post-harvest lifelong genotypes.

»   Agricultural practices that result in a product that can be stored for a long time; and integrated crop management systems.

»   Maintaining the quality and safety of fresh product by the application of appropriate post-harvest management procedures.

Less than 5 percent of agricultural research and extension funding globally goes to activities that maintain product quality and safety during post-harvest handling even though it is more sustainable to reduce post-harvest losses of already produced food. Reduced post-harvest losses of fragile horticultural crops are impossible without altering this condition.

Quality Factors

The degree of perfection or superiority that any commodity has in terms of its intended purpose is determined by a combination of qualities, features, or characteristics. It depends on the commodity and the person or market (producer, consumer, and handler) doing the quality assessment on how important a certain quality trait is. Among the most significant quality qualities for farmers are high yields, pleasing appearance, ease of harvest, and the ability of the crop to resist long-distance shipping to markets. Wholesale and retail marketers place a high value on product attributes like appearance, hardness, and shelf life. On the other hand, consumers base their initial evaluation of fresh fruits, ornamentals, and vegetables on their look (which includes "freshness"). The consumer's pleasure in terms of the flavour (eating) quality of the edible section of the produce is the driving force behind subsequent purchases. The following is a breakdown of the factors that influence the quality of fresh fruits and vegetables:

Appearance (Visual) Quality Factors

There are a variety of characteristics to consider, including size, shape, colour, shine, and the absence of flaws and deterioration. Insects, illnesses, birds, and hail; chemical injuries; and other imperfections can cause defects before harvest (such as scars, scabs, rind staining). Any of these post-harvest flaws could be a symptom of a pathological or physiological condition.

Textural (Feel) Quality Factors

Depending on the commodity, these qualities can include firmness, crispness, juiciness, and hardness. For horticultural crops, the quality of their texture is just as significant as their eating and cooking qualities. Because of their delicate nature, soft fruits cannot be delivered across great distances without considerable damage. Soft fruit shipments frequently demand harvesting at an immature stage in order to preserve the best possible flavour.

Flavor (Eating) Quality Factors

In addition to the sweet and salty flavors mentioned above, there are also fragrance and off-flavors that can be found in a product. A wide variety of chemicals are considered while determining a flavor's quality. In order to obtain relevant and meaningful information regarding the flavor quality of fresh fruits and vegetables, objective analytical determinations of essential components must be combined with subjective evaluations by a tasting panel. This method can be used to establish an acceptable threshold for a certain condition or situation. A representative sample of consumers must be subjected to extensive testing in order to determine whether a certain commodity's flavour is preferred by consumers.

Nutritional Quality Factors

Vitamin C, Vitamin A, Vitamin B, thiamine, and niacin are all found in fresh fruits and vegetables, as well as minerals and dietary fiber. Phyto-chemicals and carotenoids, among others, may help prevent cancer and other diseases by their presence in fresh fruits and vegetables (phyto-nutrients).

A commodity's usage and value are based on its quality features, which are defined by grade standards. With adequate enforcement of these standards, they are crucial for quality assurance during marketing and serve as a common language for trade among farmers, handlers and producers as well as receivers at terminal markets.

Safety Factors

The safety of fruits and vegetables is under threat from a number of sources. There is a variety of naturally occurring toxins to consider, such as glycol alkaloids in potatoes, as well as contaminants such as mycotoxins and bacterial toxins, as well as heavy metals (cadmium, lead, and mercury). While health officials and scientists place microbial contamination at the top of their list of concerns, many customers place pesticide residues at the top of their list.

Raw fruits and vegetables should typically be free of the majority of human and animal enteric pathogens unless they have been fertilized or irrigated with water that contains such waste. To avoid contamination of fresh food with pathogens like Salmonella and Listeria, organic fertilizers like chicken dung should be sterilized before use in fruit and vegetable cultivation. Products that come into contact with soil have a higher risk of contamination than those that do not. Fruit and vegetable safety is best achieved by reducing the risk of contamination during their growing, harvesting, harvesting, handling and packaging and storing. In order to reduce the risk of food borne microbial contamination, strict adherence to Good Agricultural Practices and Good Hygienic Practices is strongly recommended. To reduce microbiological contamination in the food-service, retail, and consumer sectors, it is highly advised that all raw products be handled and washed with care, and that proper hygienic precautions be strictly followed.

Factors Influencing Quality and Safety of Horticultural Crops

Genetic factors

Variation in composition, quality, and post-harvest life potential can be found within each commodity grouping. Carotenoid content and vitamin A content have been successfully selected by plant breeders for carrot, sweet potato and tomato cultivars, as have shelf-life-extended onion and tomato varieties, as have varieties of sweet corn that retain their sweetness longer after harvest, as have sugar-rich, firmer-fleshed cantaloupe and watermelon varieties, as well as pineapple cultivars with higher concentrations of ascorbic acid, carotenoids and sugars. To name a few, genetic engineering has improved the nutritional value of fruits and vegetables. There are several commercial cultivars that are chosen because of their capacity to survive the marketing and distribution but lack adequate sensory quality, particularly flavour.

Breeders of horticult t in numerous ways. Priority objectives in this regard include:

1. Maintaining high levels of flavor and nutritional value in order to meet the needs of end users

2. Reducing the use of chemicals on fruits and vegetables by introducing resistance to physiological problems and decay-causing microorganisms.

Climatic condition

Fruit and vegetable nutritional quality is highly influenced by temperature and light intensity. This means that ascorbic acid, carotene, riboflavin, thiamine, and flavonoid content can vary depending on the region of production and the season in which the plants are cultivated. The ascorbic acid concentration in plant tissues decreases with decreasing light intensity. Plants' mineral nutrient intake and metabolism are influenced by temperature, as transpiration rates rise with temperature. Harvesting and handling of harvested plant parts can be harmed by mechanical damage and decay due to variations in the amount of water available to the plant as a result of rainfall.

Cultural practice

Plant water and nutrient supply can be affected by soil type, rootstock used for fruit tree cultivation, mulching, irrigation, and fertilization. This in turn can impact the nutritional quality of the harvested plant component. Genetics and climate have a greater impact on plant vitamin content than fertilizer application. A plant's ability to absorb minerals and elements from fertilizers is significant and varies widely.

Organosulfur compounds in *Allium* and *Brassica species*, for example, are influenced by the intake of selenium and sulphur. Vitamin C has been demonstrated to increase fruit firmness and shelf life, as well as minimize the incidence of decay and physiological problems, all of which contribute to a longer fruit's shelf life after harvest. As a result of greater vulnerability to microbial damage, physiological disturbances, and decay, high nitrogen content is frequently associated with lower post-harvest life. By increasing fertilizer inputs of nitrogen and/or phosphorus, the acidity and ascorbic acid content of citrus fruits is lowered, but fertilizer inputs of potassium are accompanied by an increase in these two characteristics. Mineral deficiency can cause a wide range of health problems. Citrus rind rot, cork spot, and the red blotch on the rind of apples and pears are all symptoms of a calcium deficit that can be traced back to the fruits themselves. Corking of apples, apricots, and pears; bumpy rind of citrus fruits; and cracking of apricots are all symptoms of a lack of boron in the soil. Iron and/or zinc deficiency may contribute to the dull hue of stone fruits. Soluble solids content increases when salt and/or chloride levels are elevated. Peaches turn leathery and rough when exposed to high water stress, and nuts' kernels fail to fully grow because of sunburn. Water stress causes a reduction in fruit size and an increase in the quantity of soluble solids, acidity, and ascorbic acid. When plants receive too much water, they fracture their fruits, become turgid, become more vulnerable to physical injury, lose firmness, mature later, and lose soluble substances.

Pruning and thinning are cultural methods that affect crop load and fruit size, both of which can have an impact on the nutritional makeup of fruit. Fruit composition is not directly influenced by the use of pesticides and growth regulators, but may be affected by delayed or hastened fruit maturation. Pre-harvest disease control has a significant impact on post-harvest fruit and vegetable handling disease incidence and severity.

Maturity at Harvest in Relation to Quality

The most essential factor in determining fruit quality and storage life is the fruit's ripeness at harvest. If the fruit is too young to ripen, it is more prone to shriveling and mechanical damage and has a poorer flavour. Soon after harvesting, overripe fruit is likely to be soft and mealy with an insipid flavour. Prior to or after harvest, fruit that has been plucked prematurely or late is more prone to post-harvest physiological problems.

Fruits that need to be shipped over great distances may be collected at a mature but unripe stage of development. These fruits' maturity indices are a result of a compromise between the indices that would provide the highest eating quality for consumers and the indices that would provide marketing flexibility. Fruit can be divided into two groups:

1. Those that are in capable of continuing their ripening process on removed from the plant.

2. Those that can be harvested at mature tag and allowed to ripen off the plant.

Cane berries, cherry, citrus fruits, grape, litchi, pineapple, pomegranate, strawberry are all in Group 1 of fruits. Other fruits in this group include avocados (banana and cherimoya), bananas (cherimoya), mangoes (nectarine and mango), kiwis (kiwifruit and mango), papayas (mango and papaya), plums (persimmon and plum), and sapodillas

It is recommended that fruits of the Group 1 category be chosen when they are fully ripe to ensure the best flavour quality, as they produce very little ethylene and do not respond to ethylene treatment except in terms of de-greening (the elimination of chlorophyll). When exposed to ethylene, fruit in the Group 2 group ripens faster and more uniformly because of the higher amounts of ethylene it produces during the ripening process.

It is common for many vegetables, particularly leafy vegetables, and fruit-vegetables (like okra and cucumbers) to be at their best before they are fully ripe. As a result, harvests are frequently delayed, resulting in lower-quality product.

Method of harvesting in relation to physical damage and uniformity of maturity

the composition and post-harvest quality of fruits and vegetables can be considerably influenced by the harvesting process. Injuries caused by mechanical means increase the body's vulnerability to decay-causing infections by increasing the pace at which water and vitamin C are lost. All flowers and the majority of fresh produce are picked by hand. Mechanical harvesting is used for root crops and some commodities that will be processed. Produce quality can be greatly affected by how harvesting activities are managed, whether by hand or machine. Harvesting at the right time in relation to the product's development and weather conditions, as well as training and monitoring of personnel are all important aspects of proper management. In order to successfully handle fruits and vegetables after harvest, it is critical that they are handled quickly and carefully, cooled immediately after harvest, and kept at a temperature that prevents decay while in transit and storage.

All of these aspects must be considered, regardless of the harvesting method used. In the case of mechanically harvested goods, these aspects are much more crucial than usual.

Packing in the field just after harvest can drastically cut down on post-marketing handling time. Those fruits and vegetables that don't need to be washed before they're ready to be sold are packed in mobile field packing stations that have adequate shade.

Post-Harvest Management Procedures

Packing and Packaging of Fruits and vegetables

The packing house or the field might be used to prepare the produce for market. In addition to sanitizing and classifying the items into several categories (quality and size), waxing and applying a certified fungicide are also required before shipping. In the course of marketing, packaging protects produce from both mechanical damage and contamination. Containers made of corrugated fiberboard are widely used to package food; however re-usable plastic containers can also be utilized in this situation. It is possible to utilize packaging accessories like trays and cups to keep the product in the packaging container while still allowing moisture to be retained, chemicals to be applied, and the absorption of ethylene. Mechanical or manual packaging technologies might be employed. While in storage or transit, the temperature and relative humidity of product can be affected significantly by the packing and packaging methods used.

Temperature and Relative Humidity Management

The environment's temperature is the most significant influence on the degradation of harvested goods. The ideal storage temperature for most perishable horticulture products is around O°C. A 10°C increase in temperature, on the other hand, raises the deteriorating rate of perishables by two to three times. Internal and external elements are greatly influenced by temperature, which also has a substantial impact on spore germination and pathogen proliferation.

There are several conditions that might lead to the quick deterioration of fresh products, such as

Freezing: Perishable goods have big, highly vacuolated cells and high water content. Their tissues have a rather high freezing point (ranging from -3°C to -0.5°C), *and when* they are disrupted by freezing, their tissues collapse and their cellular integrity is lost. Refrigerator design flaws or thermostat failure can cause freezing in cold storage systems. When produce is left on unprotected transit docks during the winter, it might freeze. Storage at low temperatures that are substantially over their freezing points but below a critical point known as their chilling threshold or lowest safe point can cause harm to certain commodities. Damage from chilling includes surface and internal discoloration, pitting, water soaking and failure to mature or uneven maturity, development of unpleasant flavours and an increased vulnerability to disease assault, to name a few of the most common signs.

Heat injury: Perishable crops can also be damaged by high temperatures. Transpiration is essential for growing plants to sustain their appropriate growth temperatures. In contrast to the protective effects of transpiration, organs removed from the plant are vulnerable to localized bleaching, necrosis (sunburn or sunscald), or widespread collapse if exposed to direct sources of heat, such as sunshine.

Relative humidity (RH): It is defined as a percentage of the amount of water vapour that may be held in the atmosphere without condensation at a given temperature and pressure. Relative humidity (RH) The air's ability to keep moisture gets better as it gets hotter. Commodity and environment vapour pressure differences (VPDs) are directly related to the amount of water lost. Inversely, the air's RH has an effect on the product's VPD. Some physiological problems, water loss, rotting, and fruit ripening consistency can all be affected by RH. The RH of the surrounding air may not be as crucial in promoting degradation as long-term condensation of moisture on the commodity (sweating). Fruits should be stored with a relative humidity (RH) of 85-95%, whereas most vegetables should be stored with a RH of 90-98%.

70 to 75 percent RH is ideal for dry onions and pumpkins. It is recommended to keep some root vegetables at 95 to 100 percent RH, such as parsnips and radishes.

RH can be controlled by one or more of the following procedures:

» Adding moisture to the air through the use of humidifiers (water mist or spray, steam).

» Adjusting the cold storage room's air flow and ventilation to match the amount of produce being stored there.

» Keeping the refrigeration coils in the storage space or transport vehicle at a temperature that is within 1 degree Celsius of the ambient temperature.

» Insulating the walls of storage rooms and transit vehicles with moisture barriers.

» Lining packing containers with polyethylene and packaging items with perforated polymeric sheets.

» Damp floors in warehouses and storage facilities.

» The practice of packing non-harmful goods with crushed ice in transport containers or shop displays.

» Spraying leafy crops, cool-season root vegetables, and immature fruit vegetables with sanitized, clean water while they are being sold at retail.

Cooling methods

For fresh horticulture commodities, temperature control is the most effective strategy for shelf life extension. Crushed or flaked ice can be used to cool fresh fruit quickly and to maintain a high relative humidity (RH) throughout subsequent handling. There are some foods that aren't suitable for direct ice contact and must be stored in moisture-resistant containers. Hydro-cooling (shower or immersion systems) of goods that can withstand water contact and are packaged in moisture-resistant containers employs clean, sanitized water as a cooling medium. Leafy vegetables that emit water vapour quickly can be swiftly chilled using vacuum cooling. Refrigerated air is driven through vegetables packaged in boxes or pallet bins during forced-air cooling. Most horticultural perishables can benefit from forced-air cooling. Fresh fruits and vegetables require precise temperature and RH control throughout chilling and storage to ensure the best possible conditions. Cooling and storage facilities are increasingly relying on PTM instruments, such as time-temperature monitors, to better manage temperature.

Refrigerated transport and storage

The design, construction, and equipment of cold storage facilities must be adequate. A vapour barrier on the warm side of the insulation; sturdy floors; adequate and well-positioned loading and unloading doors; effective distribution of refrigerated air; sensitive and properly located controls; refrigerated coil surfaces designed to adequately minimize differences between the coil and air temperatures; and adequate capacity for expected needs are all important considerations in the design of the insulation. Proper air circulation should be ensured in the cold room or refrigerated truck by stacking goods on pallets with air spaces between the pallets and the walls. If sufficient cooling is to be performed, storage chambers should not be overloaded. Rather than a:ir temperature, it is better to measure commodity temperature at these facilities.

When transporting fresh produce across long distances, it is essential to keep it at a constant temperature. Air must be allowed to circulate around the product in order to allow it to cool and to disperse heat from the atmosphere and the road. Additionally, the stacking of loads must take into account the possibility of mechanical damage. Prior to loading fresh food, transit vehicles must be chilled. Transport truck loading should not be held up while harvest is cooling. The handling system should be maintained at the proper temperature. Transport vehicles should aim to minimize their impact on the environment as much as possible. The use of ethylene to kick-start ripening in transit is technically possible, and it has been done commercially on mature green bananas and tomatoes to a limited extent. Cool the produce prior to loading and ensure that the palletized product is loaded with an air space between it and vehicle walls so that temperature may be controlled. Vibration during shipping should be kept to a minimum in order to avoid bruising, which can lead to structural damage. Non-chemical insect control for markets with quarantine limitations against pests endemic to exporting nations and markets that do not want their produce exposed to chemical fumigants should be observed wherever possible by following controlled-atmosphere and precise temperature management.

Chilling-sensitive and non-chilling sensitive goods are frequently mixed in one shipment, and the best temperature and air composition for transporting these two types of produce typically necessitates compromises. Ethane scrubbers can be employed to remove this gas from the vehicle's internal circulation in the latter situation. In order to protect goods that are sensitive to cooling when carried with goods that are not sensitive to cooling at temperatures below their threshold cooling temperatures, a variety of insulating pallet covers are available.

Cold chain

In order to keep food and other perishable items fresh, they must travel through a series of stages and processes known as the cold chain. A chain is only as strong as its weakest link, and the cold chain is no exception. Due to either a lack of refrigeration, handling, storage, or humidity control, the cold chain suffers from poor temperature management. Cold chain infrastructure investment reduces fresh produce losses and quality degradation, which in turn results in net positive economic benefits.

Relative humidity management

Inadequate cold chain management can lead to losses in both profits and horticulture crops, whether because of refrigeration restrictions, incorrect handling and storage, or inadequate humidity control. It's imperative to strengthen processes, operations, and handling throughout the supply chain to overcome these shortcomings. Often, the amount of money needed to fix these problems is far less than the amount of money lost over time.

According to research conducted by the University of California, strawberries that were left to cool for an additional hour after harvest lost 10% of their value owing to decay. By increasing the frequency of delivery of strawberries to the chilling facility and instituting forced-air cooling, this resulted in an economic loss that was more than the cost of expediting the processing of strawberries. According to a University of Georgia research, a loss of US$172.50 in net income per truckload of 900 cartons was caused by improper temperature control for lettuce.

Treatments to Reduce Microbial Contamination

It has become and continues to be the primary worry of the fresh produce sector over the past few years. In October 1998, the FDA produced a "Guide to Minimize Microbial Food Safety Hazards for Fresh Fruits and Vegetables." It is founded on the following tenets:

» When possible, it is preferable to prevent microbial contamination of fresh product than to rely solely on corrective measures after contamination has occurred.

» Good agricultural and management practices should be used in regions over which producers, packers, or shippers control to minimize microbiological food safety threats in fresh product.

» At any point in the farm-to-table food chain, fresh product might be microbiologically compromised. Microbial contamination of fresh produce is caused by human and/or animal faces.

» There are four factors that determine whether or not vegetables can be contaminated by water. Microbial contamination from water used with fresh fruits and vegetables must be minimized; this is a critical issue.

» As a precautionary measure, the use of animal manure or municipal bios lid wastes as fertilizers should be closely monitored.

» Employee hygiene and sanitation procedures have a crucial role in reducing the risk of contamination of fresh fruit throughout harvesting, sorting, packing, and transportation.

It was issued in November 2002 by the US Food and Drug Administration (FDA) with the goal of giving scientific and practical knowledge on the safe production, handling, storage, and transportation of fresh fruits and vegetables to trainers, as well as to the general public.

Sanitized water must be used to prevent the spread of infections between water and produce, between healthy and sick produce within a single batch, and between batches of produce over time. These microorganisms, including post-harvest disease agents, can be quickly acquired and absorbed by plant surfaces. Points of entrance and harbors for microorganisms can be found in natural plant surface contours, natural apertures, harvest wounds and trimmings, and scuffing. After harvest water sanitizing treatments, microorganisms are mainly unaffected when they are positioned in these protected areas. This means that the sanitizer concentration must be high enough to destroy germs before they attach to or become internalized in produce. Water-related post-harvest operations, including as washing, cooling, water-mediated transport (flumes), and drenching with calcium chloride or other chemicals, all require a high concentration of sanitizer.

Treatments to minimize water loss

Newly harvested crops suffer greatly from the drying out of their tissues due to the loss of water through evaporation, known as transpiration. Transpiration water loss not only causes quantitative losses, but also affects the look, texture, and nutritional quality of the plant. Either by applying post-harvest treatments directly to the crop or by altering the surrounding environment, one can regulates transpiration.

Fruits and vegetables can be treated to reduce water loss, such as:

» Garlic, onion, sweet potato, and other root vegetables can be preserved by curing them.

» Apples, citrus fruits, nectarines, peaches, plums, pomegranates and tomatoes are all examples of commodities that benefit from waxing and other surface coatings.

» Moisture-resistant polymeric film packaging.

» Water loss from vegetables is exacerbated if it is handled roughly.

» Water is added to products that can be misted, including green vegetables.

Treatments to reduce ethylene damage

By increasing the senescence in harvested horticulture crops to 1 or more parts per million (ppm), the deterioration and post-harvest life of these crops are accelerated. To reduce the quality of leafy, floral, and immature fruit-vegetables as well as foliage ornamentals, Ethylene promotes chlorophyll degradation and induces yellowing of green tissues. Leaf and flower abscission, fruit softening, and other physiological abnormalities are all caused by ethylene exposure. Some fruits' degradation may be accelerated by ethylene's acceleration of senescence and softening and its inhibition of antifungal compound synthesis in the host tissue. For example, *Botrytis cineria* on strawberries and *Penicillium italicum* on oranges may be encouraged to flourish by ethylene in specific instances.

Temperature, exposure time, and ethylene concentration all affect the frequency and severity of degradation symptoms generated by ethylene. Either 1 ppm ethylene over 2 days or 5 ppm ethylene over 1/2 day at 10°C might cause yellowing of cucumbers. The effects of ethylene continue to build up after harvest. An ethylene action inhibitor, 1-methylcylopropene (1-MCP), protects ornamental crops from ethylene damage. Commercial use of this product at concentrations of up to 1 ppm on fruits and vegetables such as apples, apricots, avocado, kiwifruit and mangoes has been allowed by the United States Environmental Protection Agency for use on such fruits and vegetables. Many additional fruits and vegetables, as well as areas around the world, are likely to benefit from the usage of 1-MCP.

Treatments for decay control

There are numerous bacteria that attack and destroy perishable crops, which is a major source of crop losses. At any point, the plant organ can get infected with fungus or bacterial pathogens. Once the fruit begins to ripen, "latent" infections, in which fungus have already infected tissues, become visible.

Post-harvest rots are produced by a diverse variety of microorganisms, and they are typically induced by hard handling during the marketing process. Despite its grey colour, the *Botrytis cineria* grey mould is a major source of crop losses (such as grapes, kiwifruit, pomegranates, raspberries, and strawberries). Most of the time, viral contamination degrades the appearance of perishable goods, but viruses can also alter the flavour or composition of food items.

Certain morphological and physiological changes that can extend the storage life of some root crops can be facilitated by post-harvest curing. Root, tuber, and bulb crops lose water and degrade more quickly when stored in this manner; therefore it's a simple but highly effective solution.

Sanitization procedures include treatment to minimize microbe populations on equipment, the commodity, and the wash water used to clean it. To remove the nutrients that allow bacteria to thrive on the surface of fruits and vegetables, water washes alone are effective. Degradation-causing organism inoculum from fruit surfaces can be reduced by spraying or dipping water with a sanitizer, which also inactivates spores suspended in solution from fruit or soil and stops their secondary dissemination in water. In order to prevent the development of anthracnose, mangoes can be dipped in hot water for five minutes at a temperature of 50 °C in order to reduce the likelihood of the disease returning. Other treatments include the use of post-harvest fungicides, as well as biological control agents, such as "Bio-Save" (*Pseudomonas syringae*) and "Aspire" (*Candida oleophila*).

Treatments for insect control

During post-harvest handling, fresh fruits, vegetables, and flowers may be infested with a high number of insects. Many of these insects, particularly the fruit flies of the family, are harmful to the environment. Trade between countries can be severely disrupted by Tephritidae (e.g. Mediterranean fruit fly, Oriental fruit fly, Mexican fruit fly, Caribbean fruit fly). Disinfestations methods like irradiation, which have been found to be effective, will make it easier to trade in fresh fruit around the world. There are several factors to consider when determining which disinfestations treatment is best for a certain commodity, including price, effectiveness against pests of concern, safety, and the capacity to retain and maintain the quality of the product being treated. As an alternative to radiation, currently allowed quarantine treatments include certification of insect-free zones, chemical use (such as methyl bromide, phosphine, and hydrogen cyanide), cold therapy and heat treatment, as well as combinations of these treatments. To combat the spread of disease, researchers are currently experimenting with the use of fumigants (carbonyl sulphide and/or methyl-iodide) and insecticidal atmospheres (oxygen concentrations of less than 0.5 percent and/or carbon dioxide concentrations ranging between 40 and 60 percent) alone or in combination with heat treatments, and ultraviolet radiation. These therapies, on the other hand, are not all-inclusive and may be phytotoxic if used on certain crops.

When irradiated with levels of 50 to 750 Gy, most insects become infertile. Insect sterility is dependent on the species and stage of development of the insect in question. Because of its effectiveness in stopping the reproduction of tropical fruit flies, an irradiation dose of 250 Gy was allowed by US quarantine authorities for use on fresh commodities such as lychees, mangoes, and papayas. Most fresh fruits and vegetables may withstand irradiation levels of 250 Gy with minor quality degradation. Some commodities, however, can be damaged by doses between 250 and 1000 Gy. Compared to non-fruit crops and cut flowers, fruits are more resistant to the projected dose range. Irradiation can cause fresh produce to lose its green colour (yellowing), to drop its leaves and petals, to discolour, and to ripen in an uneven fashion. These are all side effects of irradiation. It's possible that the negative consequences won't be apparent until after the product hits the market. In order to ensure that the irradiation treatment is safe and effective, it must first be evaluated on particular products.

Enhancing Quality

Modified atmosphere storage

Extending the post-harvest life of fresh horticultural perishables can be accomplished through the use of controlled atmospheres (CA) or modified atmospheres (MA) in addition to maintaining the optimal temperature and relative humidity ranges. Delaying ripening, preventing senescence by retarding the growth of decay-causing pathogens, and controlling insects are all benefits of optimum oxygen and carbon dioxide concentrations. A commodity's susceptibility to decay can be increased by placing it in CA conditions that aren't designed for it. In recent years, CA storage technology has seen a number of improvements. Low (0.7 to 1.5 percent) oxygen concentrations that can be accurately monitored and controlled; rapid CA establishment, ethylene-free CA, programmed CA (such as storage in 1 percent 02 for 2 to 6 weeks followed by storage in 2-3 percent 02 for the remainder of the storage period) and D. The advancements in science and technology continue to benefit the usage of CA in refrigerated sea containers. If commodities (like apples and kiwifruits) had been harvested and stored immediately in CA, then CA transport is used to continue the chain. Transport via CA enables harvesting of bananas at an earlier stage of maturity, resulting in higher yields in the fields. Avocados shipped via CA can be shipped at a lower shipping temperature (5°C) than if they were shipped via air, because CA reduces the symptoms of chilling injury. Pest control without the use of chemicals is now possible in commodities destined for markets with restrictions on pest's native to exporting countries and for markets that value organic produce, thanks to CA and precision temperature management.

At both the pallet and shipping container levels, the use of polymeric films for packaging produce and their use in MAP (modified atmosphere packaging) systems continue to rise. Extending the shelf life of fresh-cut fruits and vegetables is a common application of MAP (usually designed to maintain 2-5 percent O_2 levels and 8-12 percent CO_2 levels). Use of MAP absorbers that include ethylene, CO_2, oxygen, and water vapour is on the rise. Research into using coatings to alter the interior environment of commodities has been extensive, but practical applications are still limited by biological variability. When it comes to apples and pears, CA is most commonly used for storage and transportation. Kiwifruits, avocados, persimmons, pomegranates, nuts, and dried fruits are some of the other fruits to which it is applied. It is used to transport apples, avocados, bananas, blueberries, cherry figs, kiwi-fruit mangoes kiwi-fruit nectarines peaches pears plums raspberries and strawberries long-distance via the atmosphere. A positive benefit-to-cost ratio is essential if CA is to be applied to fresh fruits and vegetables in a more widespread fashion. A commercial application of MA and CA has been limited by their relatively high costs, despite their effectiveness in extending post-harvest life for many commodities. A positive return on investment (cost/benefit ratio) can be shown in only a few instances. When strawberries shipped in air were compared to those shipped in an environment with 15% CO_2 enriched air (modified atmosphere within pallet cover), losses from decay were found to be 50% lower when shipped in the modified atmosphere (an average of 20 percent losses was sustained in strawberries shipped in air vs 10 percent losses in those shipped by MA). Because of this, the cost of using MA (US$15-$25 per pallet) was much higher than the economic loss of 10% in value.

Marine transportation using CA can extend the post-harvest life of fruits and vegetables that would otherwise have a short post-harvest potential, allowing the use of marine transportation instead of air transportation for the shipment of such products. Savings achieved by using marine transportation far outweigh the additional costs incurred by CA service.

Ethylene Exclusion and Removal

Most horticultural products, including many green vegetables, are highly susceptible to ethylene damage. Because of this, their exposure to ethylene should be minimized. There are several ways to reduce the risk of contamination in ripening rooms, including.

» Using ethylene levels of 100ppm instead of the higher levels commonly used in commercial operations.

» Venting ripening rooms out to the outside after exposure to it,.

» At least once per day ventilating the area around the rooms or installing an Ethylene Scrubber.

» The use of battery-powered forklifts instead of engine-driven units in ripening areas is also recommended.

Commodities that produce ethylene should not be stored or transported with commodities that are sensitive to ethylene. Ethylene can be removed from the air by using potassium permanganate, an ethylene oxidizer. Some commercial storage facilities use catalytic ethylene oxidation scrubbers to a limited extent.

Return on investment in reducing ethylene damage

Ethylene scrubbers have been found to diminish russet spotting in lettuce storage facilities, according to a University of California research the ethylene scrubber paid for itself many times over in the increase in lettuce's worth as a result of the difference between lettuce that was shielded from ethylene and lettuce that was exposed to it. Ethylene levels as low as 50 ppb have been shown to cause kiwi fruits to soften fast.

Criteria for the selection of appropriate post-harvest technologies

No matter how a product is distributed, the essential guidelines for post-harvest quality and safety preservation remain the same (direct marketing, local marketing, export marketing). But the level of technology required to meet the ideal conditions varies depending on the distance and time between the production and consuming sites, the planned use of the produce (fresh vs. processed) and the target market.

Fresh market fruits and vegetables can be assured of their quality and safety by following good sanitary methods and harvesting with care, even if the point of sale is just a few hours away from the harvest site. Refrigeration and packing are crucial for transporting fresh products over a long distance. When considering post-harvest technology, the following should be taken into account:

a.   The technology utilized elsewhere may not be the best for application in a developing country's specific circumstances. B) Efficiencies in labour, material, and energy consumption, as well as environmental protection, have driven much recent advancement in post-harvest technologies in industrialized countries. Only those techniques that are appropriate for the local environment should be taken and employed from other countries.

b.   Without effective management, even the most expensive equipment and facilities are of little use. Furthermore, if consumers in the target market cannot handle the additional expenses of handling facilities, an overinvestment could lead to financial losses. More important than the sophistication of the technology used in post-harvest handling is the proper education of all stakeholders throughout the post-harvest chain. To ensure the quality and safety of a product or service, it is essential that employees are properly trained and monitored.

c.   In many circumstances, commodity requirements can be addressed by using simple and affordable ways. Procedures for ensuring the proper temperature of produce include: (1) Protecting crops from the sun. (2) Harvesting at cooler times of the day or even night. (3) Ensuring that containers and non-refrigerated transport vehicles have adequate ventilation. (4) Evaporative cooling or night ambient air can be used to cool produce.

d.   The quality and quantity of fresh horticultural commodities are greatly affected by mechanical damage that occurs in all handling systems. In order to reduce the risk of mechanical injury, harvesting and handling should be simplified and workers should be educated about the need of safe handling.

e.   When it comes to successfully marketing produce, food safety throughout the post-harvest handling system should take precedence over any other consideration.

Research and extension staff in a given country must work together to solve post-harvest technology issues. As a result, post-harvest horticulturists must work in tandem with other experts in the field, including production horticulturists, agricultural marketing economists, engineers, food technologists, microbiologists, and others. The majority of post-harvest handling system problems can be solved using current information rather than conducting fresh research. Post-harvest handling systems in underdeveloped countries could benefit from the following:

1. Examine how losses in quality and quantity occur during harvesting and post-harvest handling of main commodities.

2. For each commodity, determine which tools and facilities are available in that region for harvesting and packaging as well as transportation, storage, and marketing during that season.

3. Evaluate the impact of simple handling system changes (such as harvesting method, container type, and quality sorting) on quality and safety maintenance in the system.

4. Make sure everyone who can benefit from harvesting and handling practices is aware of them. In order to reach the intended audience, all appropriate extension techniques should be employed, regardless of their complexity.

5. Determine what needs to be done to address issues that demand further investigation and then carry out that research.

Chapter - 7

Processing of
Fruits and Vegetables

As a scientific and technological endeavor, food processing encompasses more than just food preparation and cooking. It entails using scientific principles to slow down food deterioration caused by microorganisms, food enzymes, or environmental variables like heat, moisture, and sunlight - and so preserve the food. A lot of this knowledge is conventionally understood and used through experience and knowledge that has been passed down through the years. Food processing is typically a source of employment and household income in poor nations. In these conditions, producers must contend with both domestic rivals and foreign imports. With a few notable exceptions, older processing techniques typically result in food that cannot compete with "newer" items. Regarding the packaging and presentation of the processed foods, this is very crucial.

Following are some general advantages of small-scale fruit and vegetable processing for individuals in underdeveloped countries:

There are plenty of raw resources available (often in surplus).

- » The majority of technologies are readily available, usable, and reasonably priced at sizes appropriate for small companies.

- » Products, if chosen wisely, can frequently be produced locally, adding to employment and meeting a demand that is high.

- » Small-scale fruit and vegetable processing is particularly suited to women when compared to some other technologies.

»　Most procedures have minimal effects on the environment.

However, choosing the right items for small-scale manufacturing, as well as the techniques to make them, requires critical thought. Many "advisers" make the mistake of assuming that just because a raw material is in excess each year, a successful fruit and vegetable processing business may be started to utilize the surplus. Before a firm is established, there must be a need for the processed food that is clearly determined. If not, the most likely outcome is the production of a processed good that no one wants to buy, resulting in significant financial losses for all parties.

There are several examples of misdirected development projects that have arisen from a desire to treat rotten fruit in order to avoid mounds from amassing each season. Instead, shelves of more costly, decaying processed meals that nobody wants to eat are the outcome.

Generally speaking, the products that are best suited for small-scale production are those for which there is a large demand and for which processing can add more value. Fruits and vegetables often cost less when they are raw, but they can be processed into a variety of snack foods, dried foods, juices, pickles, chutneys, etc. that are much more valuable.

Because of the high added value, only a small amount of food needs to be processed in order to generate a respectable income. As a result, the size and type of equipment needed to operate on this scale may be kept at prices that most prospective business owners can afford.

The safety of the generated fruit and vegetable items is another factor that is relevant to both domestic and industrial production. Despite the fact that fruit and vegetables share a lot of similarities, it's crucial to recognize the following differences: The majority of fruits are acidic, and this acidity regulates the kinds of microorganisms that can flourish in the processed goods. While moulds and yeasts are able to develop and generate clear visible spoiling, food-poisoning bacteria cannot under the acidic conditions, preventing customers from ingesting the food. Yeasts and moulds infrequently result in severe food poisoning, even if a contaminated fruit product is consumed.

However, vegetables tend to be less acidic than fruits, and a larger variety of microorganisms—including germs that can cause food poisoning—can flourish in vegetable products. This is especially risky when specific bacteria transfer toxins (or poisons) into the food yet don't cause immediate signs of deterioration. Customers might so eat the poisoned food without being aware of the contaminated germs.

As long as the food is dried properly and kept dry, microorganisms cannot thrive on dried fruits and vegetables. However, before the food is consumed, these microorganisms may reappear if it was extensively polluted before drying or if it was allowed to become damp during storage.

Natural enzymes, in addition to the operations of microorganisms, swiftly alter the colour, flavour and texture of fruits and vegetables after harvest. Microorganisms and enzymes, in general, have a characteristically short shelf life and fast processing after harvest is consequently required.

Vegetables and fruits are widely available raw materials in many countries. Vegetables have traditionally been grown in tiny garden plots in most tropical regions, while fruit trees have traditionally been planted around the house or complex to give shade, timber, and, of course, food. A wide range of home gardening and orchard planting and regeneration initiatives are emphasizing and promoting these activities, with the goal of addressing environmental issues including soil erosion and improving fruit quality and variety.

Together with an increasing awareness of the value of processing for improved food security and income generation, these programmes have led to fruit and vegetable processing being seen as an important method of improving rural and urban populations' livelihoods by numerous development agencies and government institutions.

However, for these projects to be effective, the projected benefits of processing should be clearly outlined from the outset. Participating groups or individuals may receive direct benefits, such as:

» Improved fresh produce storage with minimal loss.

» Increased health and well-being through year-round eating of fruits and vegetables.

» Preserving seasonal surpluses that might otherwise go to waste while increasing profits from the sale of processed fruits and vegetables

The increasing demand for raw materials by processors may also boost backward links to farmers and suppliers (particularly in integrated development programmes). This could be the case:

» Increase the production of fruits and vegetables and their yields, quality, and variety.

» Fruit and vegetable growing will enhance the rural environment. Such enhancements might include soil regeneration and rainwater collection as a result of planting fruit or vegetable orchards.

» Enhancing farmer incomes through increased raw material sales to processors or establishing supply agreements that lessen farmers' reliance on seasonal price volatility.

» Demand for equipment, packaging materials, and ingredients is rising, which encourages the growth of related supplier businesses and forges strategic partnerships between various manufacturing industries.

Processing for Home Consumption

Since most fruits and vegetables have short growing seasons, processing is often done to ensure a supply of these commodities when they are not available in their natural state. This is due to the fact that frequent eating of fruits and vegetables is required to maintain health, as well as the fact that some (such as tomatoes, onions, spices, etc.) are used in everyday cooking in many societies.

Fruit and vegetable eating has varying levels of importance in the diet in various communities around the world. Fruit and vegetable farming can be challenging and has a limited annual growing season in some places, such as the chilly, barren environments of the Central Asian steppes, mountainous sections of Latin America, and dry deserts in Africa and Asia. As a result, these products do not play a significant role in local diets, which place a greater emphasis on animal products. In practically all of these civilizations, however, modest quantities of leafy vegetables and harvested fruits are conserved for the winter or dry seasons and as a result of their scarcity, they have a high value due to the necessity of eating fruit and vegetable products to maintain health.

Contrarily, the soils and temperature are ideal for the growth of fruits and vegetables in practically every country in the wet tropics and subtropics, and there has historically been a reliance on these products as an essential component of the typical daily diet. Indeed, in certain places, religious convictions, such as Hinduism, which mandates vegetarianism for adherents, may encourage this. Vegetables are consumed daily in many communities across the globe as a side dish to a main cereal or root crop meal, with or without meat and fish. Fresh fruits are consumed as appetizers, snacks, drinks, and desserts while they are in season.

Many conventionally produced processed fruit and vegetable items are traditionally manufactured at home. Asian examples include fruit pastes, pickles, chutneys, and fruit "leathers," which are dried sheets of fruit paste that resemble thin leather. A vast variety of dried chips, powdered leaves, fruit beers, and other goods made from fermented fruits and vegetables are traditionally manufactured throughout Africa and Latin America. Although it is impossible to distinguish between any of these products clearly, they are all frequently designed for storage and not for sale. a few goods that are prepared for home use

When extra money is needed, consumption may also be sold to neighbors or in village markets. Due to low salaries, home processing is typically done without specialized equipment, using already-existing cooking utensils and straightforward parts that may be built from inexpensive local materials. Although some processing procedures are difficult and sophisticated and include a series of processes that must be appropriately completed in order to be effective, they can also be affordable.

Home processing serves a number of purposes, including ensuring food security during times of scarcity and preserving health by ensuring that the diet has enough nutrients. More information about these features is provided below.

Food Security, Nutrition and Health

Fibre and a variety of vitamins and minerals can be found in abundance and at a low cost in fruits and vegetables. Foods that are consumed fresh tend to have the most nutritional value, while fermented foods may have a higher B-vitamin content as a result of the fermentation process.

Fruit and vegetables, on the other hand, are highly perishable and must be processed if they are to be stored for an extended period of time. Depending on the sort of plant material that is being investigated, preservation may or may not be necessary. At room temperature (25°C), for example, bulbs and tubers can be stored without preservation for far longer than fast-growing components like shoots.

Both the type of fruit or vegetable and the processing method employed have an effect on the amount of nutrients lost.

To summaries, home processing of fruits and vegetables serves the following purposes:

» Preserving seasonal surpluses to provide food security during times of scarcity.

» Year-round supply of cooking ingredients.

» Having partially prepared goods on hand when you need those increases convenience.

» The availability of a wide variety of foods throughout the year to ensure a well-rounded diet.

An educational and training component is often important when a program's goal in improving health and nutrition status is to increase the effectiveness of food processing. It is frequently required for participants to form groups in order to benefit from one other's experiences and to make such interventions cost-effective. If there are already groups like mothers' clubs or farmers' societies in the community, they can be recruited to serve as the focus of the new programme. Health centers have been related to some of the most effective food processing innovations that have resulted in better nutrition.

If a facility or piece of equipment is to be used for a long period of time, it must be low-cost and easy to maintain. While a new structure or piece of processing equipment can be donated, if money and responsibility for maintenance are not established at the outset, the facility will not be able to function. Various sorts of producer organizations and village co-operatives have been established, with variable degrees of success, in an effort to reduce the costs of processing. The level of mutual trust and readiness to work together that exists within a community is a significant consideration. It can take years to introduce these principles if there is no history of a collaborative approach to issue solving or adopting a new development. Many examples of effective village organizations have been in existence for a long time and have supported on-going training programmes in nutrition as well as the construction of community systems and facilities for processing crops.

Improvement to Home Processing and Storage

Process and storage approaches based on conventional wisdom are more likely to be adopted successfully than more complex ones. Even while new technology may not always be suitable to the local area, improvements to established processes have a better chance of succeeding. Improved drying and storing are notable examples of this method.

Drying

As a result, drying is one of the most widely used processing methods. Food security can be significantly enhanced in most regions if it is paired with improved food distribution systems. Important examples of dried fruits and vegetables are okra, cabbage, spinach, mango slices or leathers, garlic and other flavorings, all of which are stored for use in daily cooking. When it comes to food preservation, there are few costs involved and the drying process can be shared with family members of all ages, even young children. For example, cost and workload are two of the reasons why solar dryers have not been widely adopted in rural communities. Even if the simplest drier is built from locally accessible materials, it still costs more than drying in the sun. Because children are not capable of loading and carrying trays of produce, this additional burden is placed on their parents, mostly mothers, who bear the brunt of this added work load. A family's satisfaction with their current dried food and drying time may not be influenced by the potential for greater quality goods and quicker drying times utilizing solar dryers.

Improved sun drying is expected to be more effective in these situations. Crops can be raised off the ground on woven matting or wooden drying frames to reduce contamination by dust, crawling insects, or rodents; mosquito netting can be placed over the crop to protect it from birds and flying insects; smaller pieces of fruit and vegetables can be cut to speed up drying; and blanching vegetables can help retain their colour. Some programmes may be able to link drying advances to agricultural advancements and the production of crop varieties that are better suited to drying than traditional types.

Concentration by boiling

However, in some cultures, key food items like tomato paste, extracts from wild plants like baobab fruit, chutneys and sap syrups prepared from fruit or sap are commonly consumed despite the additional fuel-wood use associated with home-preserved produce. As the boiling process nears its conclusion, it becomes increasingly thick, it is important to keep an eye on the rate at which the product is being heated in order to avoid burning it in certain areas.

Fermentation

It is not uncommon for people to make their own fruit wines and beers at home in nations where alcohol is not forbidden by religious views. A broad variety of fruits, including pineapple, melon, and pawpaw, are employed. Banana is used in the sample below, which is from East Africa. The following description by the processor demonstrates the difficulty and level of skill necessary to generate a quality product from bananas.

Fruit is picked when it is just ripe enough to eat, but not so ripe that it will spoil the rest of the crop. For up to four days, they are kept in a pit covered with banana leaves. Bananas can be warmed and ripened faster by lighting a fire in a small hole heading into the pit. It is time to peel the bananas and arrange them in a wooden basin. Walking on the fruit then breaks it down into a pulp. Two grasses, one of which is young elephant grass, are cut to a length of 15 cm and then carefully combined to the precise proportions, as seen in the picture. It's a gradual process that sees layers of grass pulped and then layered on top of each other. This is done to break down the pulp into smaller pieces, allowing the juice to be released. Due to their lack of flavour, these two grasses are the only ones that can be used. A coarse filter frame consisting of wooden poles nailed together over a trough is used to force the juice out of the pulped grass after it has been pulped for 30 to 45 minutes. Next, a calabash is used to collect the juice, with a hole cut into one side and a hollow handle added to the top. Through the hollow handle, juice is poured into a second calabash that is formed like a runnel and lined with grass as a fine filter. It is then strained through the grass. It is necessary to collect the turbid but particle-free juice in a sterile container. At this point, it is tasted and, if necessary, water is added to regulate the sugar content.

Sorghum is subsequently put at a ratio of 2 Kg per 20 liters of standard juice in a clean wooden trough. In order to keep the temperature of the fermentation at a constant 22-26°C, the trough is covered with banana leaves and the pulp residue from pulping. A calabash/grass filter is used to strain the liquor into clean containers or bottles once it has fermented. A pasteurization or clarifying step is not required; the alcohol concentration of the beer serves as a preservation mechanism (3-5%). As a result, it can be kept for up to four days at room temperature before needing refrigeration. If the product was clarified by filtration or sedimentation and pasteurized in the bottles, the process might be enhanced and the shelf life increased.' South and Central Asians are more likely to preserve fruits and vegetables by fermenting them into pickles than those in other regions of the world. Lime pickles, mango pickles, mixed fruit pickles, and pickled vegetables, such as various kinds of cabbage, are just a few examples. In India alone, there are an estimated 700 different types of fermented pickle, and each region has its own unique mix of ingredients and processes.

A natural sequence of bacteria ferments sugars in the raw ingredients to produce lactic acid in all circumstances where the prepared fruit or vegetable is mixed with spices and held in a closed container such as glazed pottery. This product can be maintained for several months while being used as long as the raw materials are not significantly infested with mould and insects that can carry yeasts or moulds on their bodies are kept out of the container.

A 2 cm layer of salt is poured in the bottom of a wide-necked earthenware drum or jar in Central Asia to prepare pickled cabbage. On top of the salt, arrange a 5 cm layer of finely shredded cabbage, then fill the jar in layers, alternately adding salt and cabbage. A wooden or cloth bung is then attached and the mixture is left to ferment for a few weeks. Pickling brine is formed when the water in the cabbage is drained away. The cabbage is weighted down with a clean stone if it is not going to be used right away to keep it submerged in the brine and prevent it from drying out and rotting. There is nothing that can be done to improve these processes without a significant increase in capital and operating expenditures. Adaptable to local tastes and environmental conditions, they are a good choice.

Pickling

There is also a wide range of unfermented pickles produced in Asia. Two examples below describe the production of lime pickle and mixed pickle from South Asia.

Washed and split into four (without separating the pieces), well-ripened limes are immersed in a salt solution that is both concentrated and not over-ripe. After that, they're let to air dry for a couple of days. Lime is often dried with salt crystals sprinkled on top of it. When the skin turns brown and breaks apart easily with your fingertips, the drying process is complete. The result is then packaged in clay containers with the drained juice. The salt removes water from the pieces due to osmosis and some sugar is also removed. As a result, salt-tolerant micro-organisms begin to grow while the product is being dried. These bacteria produce a small amount of acid, but their primary function is to impart the product's distinctive flavour, which is a sour and salty mix that may be stored for months.

The main raw materials for mixed pickle are green chillies, onion, papaya, and spices. Peel and thinly slice the papaya and onion. Additionally, the spices are rinsed before being ground. In a clean pot, combine all of the ingredients, including vinegar, and mix thoroughly by hand. The product can be kept in the sealed pot for several months, during which time there is a degree of fermentation which improves the flavour, but it does not contribute to the preservation. Dates, pineapple, and mango are some of the raw components used in other mixed pickle compositions.

Storage

Inadequate food storage methods are often the root of the most significant food security issues. Dried goods for example are vulnerable to mould and insect and rodent infestations if they aren't properly dried. It's not uncommon to suffer significant cereal losses during storage, which can be lethal. Traditional fruit and vegetable storage methods vary significantly over the world, but they all rely on locally accessible materials that cost little or nothing. Storage practices and facilities can differ greatly from place to place; however, the following sections discuss the broad concepts of storage and provide examples that may be helpful in improving storage practices and facilities in different locations.

Dried foods, fermented and pickled foods, and concentrated syrups or pastes are the most commonly preserved processed fruits and vegetables in the household. The items in the latter two categories are all moist, but they've been preserved with the addition of lactic acid, salt, or sugar. An airtight, insect- and rodent-proof seal, along with an impermeable container to confine the contents, are all that are needed for adequate storage. Glazed pottery or glass vessels, with cork or wooden stoppers, sealed with resin or wax are entirely adequate for the expected shelf life of these products. In a cool, dark spot, away from sunlight and wetness, they should be kept.

Insects, animals, and moisture are all threats to dried fruits and vegetables. Green leafy vegetables like carrots, which contain high levels of carotene, must be protected from sunlight and crushing in order to maintain their integrity. Many households choose to hang dried fruits and vegetables from rafters or above cooking areas, where they are guaranteed to keep dry, because earthenware jars can be prohibitively expensive. Most animals are deterred by this, but insects are still able to get through. As an example, a basic mesh basket made of wooden sticks and mosquito nets can be slung from the attic's rafters to improve storage space.

Only oil or kerosene can be used to keep crawling insects at bay when storing on tables or elevated platforms in the home. Insect-proof mesh should be used to cover storage containers. Once they are fully cleaned and dried, re-used cans with press-on lids or other insect- and rodent-proof containers can be used.

Home Processing to Earn Extra Family Income

Throughout the developing world, a vast number of people process their own food and sell it to neighbors or at local markets in order to supplement their families' incomes. Because of a lack of working capital or an unregulated cash flow, those in the informal sector have limited resources to invest in equipment and must acquire inputs like raw materials, ingredients, and packaging in tiny quantities from retailers. In other situations, they may be unable to compete on price with products from the official sector or even with subsidized imports. They have limited access to credit because they lack collateral and credit providers are unable to justify the costs of servicing a large number of small loans.

The lack of quality control experience among informal processors means their products cannot compete with those from the formal sector in terms of quality, safety, or aesthetic appeal. They can't create trusted connections with suppliers or retailers since they just process produce until they have enough money to meet their current demands and then stop producing until the next financial need. As a result, getting future orders may be more challenging for informal firms. As a result, if a programme aims to aid informal, home-based processors, it must consider a wide range of variables. It's important to have easy access to credit, but that's not enough to guarantee success. In addition to technical training in production procedures, hygiene, and quality assurance, a better capacity to discover appropriate markets and develop abilities for marketing the products is required. This is the most significant requirement.

For processors in the unorganized sector, several development programmes are having some success raising their wages. In order to compete with large-scale producers of fruits and vegetables, small-scale farmers and rural families typically cannot obtain all of the necessary skills. Problems like these can be avoided in some situations by forming teams that have a high level of mutual trust among its members. People benefit from each other's knowledge and experience, as well as from the group's collective purchasing power and the ability to obtain credit due to the group's higher demands.

In addition to enabling a more flexible working environment to take into account domestic duties, the division of labour within the group also has social benefits such as developing, encouraging, and rewarding an individual's competence and talents in their area of work, as well. Individuals' self-esteem and will to strive can suffer as a result. Instead of each group member having to be an expert in every aspect of running a processing company (such as quality control, supplier or customer negotiations, record-keeping, production of financial data, and marketing and sales, for example), this approach allows each member to focus on what they are best at and make a unique contribution to the group's overall success. It is possible to organize these groups in several ways: as worker cooperatives where each member has an equal stake; as religious co-ops; as family-centered organizations; or as organizations based on preexisting activities such as a young people's club, mother's union, or village development society.

The division of labour in the informal economy is a new trend in group activity. A number of organizations can agree to work together and split the revenues of a joint endeavor. Often, farmers' groups work with local processors to supply them with raw materials of a certain quality and quantity. With this, farmers are free to focus on farming instead of learning how to market their products. The raw materials' transportation expenses can be cut in half if the processors acquire from a variety of local farms. Marketing groups buy the products from processors, who focus on manufacturing high-quality products at the lowest cost. If the marketing department buys from multiple processors, they can narrow their focus to the products with the largest profit margins and most stable markets.

As this system evolves, it may include processing some fruits and vegetables in-house to save on transportation expenses while also increasing the market value for local farmers. Returnable containers can be used to transport fruit from the farm to a nearby processor, where it can be dried or crystallized. Since transportation expenses are minimized and pollution at the central processing plant is eliminated, as well as because fruit and vegetable waste can be used on the farm for green manure, this has some environmental advantages. By blending or sorting, the processor ensures a consistent quality across all batches, which are then packaged under a single brand name. In addition to saving money by purchasing packaging in bulk, this enables suppliers to meet the needs of merchants who might not otherwise consider purchasing in bigger quantities. As a result, processing businesses come in many shapes and sizes, ranging from the intermittent work of a single person from home to the more formal structures of single and group informal firms.

Processing for sale

A number of factors have contributed to a growth in developing countries' interest in small-scale food processing as a means of launching new businesses. These are some examples:

» In rural and urban regions, stronger promotion of private sector micro- and small-scale firms by national and international development organizations.

» The emergence of new culinary preferences, particularly in major urban areas.

» The ambition of national governments to boost export opportunities and foreign exchange revenues while also developing national industrial capabilities.

» Food surpluses due to effective agricultural interventions or development programmes in some countries, which necessitate storage and processing.

» A growing sense of pride in local items that can compete with imported goods.

Successful fruit and vegetable processing businesses, on the other hand, necessitate abilities that go beyond those required for home processing. In contrast to home or village processing, where customers and processors know one another and can provide feedback on the quality or perceived value of a processed item, formal food processing enterprises rarely have access to information on who is eating their food or what they think of it. Consumers are similarly cut off from the people who grew their food, with only a few avenues for feedback between processors and purchasers. As a result, in order to find clients, processors must learn new skills such as producing attractive packaging and marketing and selling.

As a result, people expect to find the same food in every pack, with the same quality, every time they purchase a product. In order to ensure the consistency and uniformity of their products, producers must take steps to improve their process control and acquire new quality assurance abilities. Aspiring entrepreneurs must also deal with legal issues, such as registering a business, paying taxes, establishing employment policies, and preparing a business plan; these are just a few examples.

Those who want to make a living primarily from fruit and vegetable processing have unique challenges that distinguish them apart from most other entrepreneurs.

» If they aren't treated soon after harvest, many raw materials will spoil quickly.

» Additionally, many fruits and vegetables are prone to seasonality, which means that the business can only run for a portion of the year unless crops are either partially processed for intermediate storage or a succession of crops is processed throughout the year.

» The cash flow of a business is also affected by the seasonality of its raw materials, which must be purchased within a relatively limited harvest time. The yields of fruits and vegetables vary greatly depending on the weather, particularly the patterns of rainfall and plant diseases.

» Because of this, raw material prices might fluctuate widely, making it more difficult to plan a business' operations.

» Some processed goods also have a seasonal demand (for festivals, rituals, etc.) that complicates business planning and cash flow even further.

» After processing, some fruits and vegetables have a limited shelf-life, and distribution systems must be organized to ensure that clients receive the correct amount of products before they expire.

» A high quality of cleanliness and manufacturing control must be maintained to avoid the possibility of contaminating or even killing clients by allowing the growth of food-poisoning microorganisms to thrive in the products.

A fruit and vegetable processing business, on the other hand, is a more complicated undertaking than a person can do at home. Many people are able to learn how to manufacture high-quality items, but this does not mean that they can then go on to become successful entrepreneurs. Many small-scale food processors begin their careers at home, but their success depends on a variety of factors, including their planning and business skills, their creative flair to produce products that have a demand and are different from those of their competitors, and their determination to succeed.

Successful entrepreneurs typically rely on food processing as their primary source of income. Some of these entrepreneurs seek out loans or arrange working capital to buy specific machinery, while others learn business and marketing skills so they can broaden their firm. Many countries and regions of the same country have various levels of service availability (electricity, clean water, gas, servicing and maintenance skills and facilities etc.). Financial resources such as cash on hand and the ability to seek loans are likewise very variable. Because of this, it is impossible, however, to provide a detailed description of the conditions under which any firm can succeed, but where applicable, examples of successful small fruit and vegetable processing enterprises are provided.

Decisions about which technologies to use are complex and interconnected, but some of the most important considerations include the following:

» Equipment and ancillary services, including the cost of purchasing and maintaining them.

» Costs of operations and overall financial success.

» Actions to ensure the health and safety of the workers.

» Meeting the requirements of an already established administrative or industrial process.

» Requirements for the operation and maintenance of a facility.

» Pollution of air or rivers in the immediate area.

This is not to say that these criteria are merely a list, but rather that each of them will have a varied weighting in certain situations. In order to make the best possible decision in a given situation, it is impossible to come up with a straightforward answer.

When launching a new business, the basic rule is to perform market research and produce a preliminary feasibility study after deciding on a product to produce. If the firm appears to be successful, create a business strategy, secure financing as needed, and register the company with the IRS and local government agencies. After that, you'll need to set up production facilities, negotiate contracts with vendors and clients, and oversee the products' routine production. Keep an eye on the competition and challenging market conditions and adjust your business strategy accordingly.

As a beginning point for entrepreneurs, the following sections describe the product characteristics and production processes for a wide range of fruit and vegetable products.

Selecting Products and Production Methods

Fried products

Starchy fruits like breadfruit, banana, and jackfruit can be fried and consumed as snacks. Enzymes and microorganisms are destroyed by heat during frying, and if the product is properly dried and stored, it can last for several weeks. The temperature of the oil used for frying must be carefully controlled, not only for safety concerns, but also for financial and quality reasons, as hot oil can splatter onto operators when wet fruit is immersed. If you overheat your oil, you'll notice that it emits an unusually blue haze. When the oil begins to thicken and have an unpleasant taste, this is a sign that the oil is breaking down chemically. To put it another way, flavours are transmitted to the final product, which is a deal-breaker. There is a larger expense in purchasing more oil than necessary when oil is thickened. Shelf life is also reduced if the product has too much oil on it.

Bottled and canned products

Both bottling and canning involve filling containers with food and heating it to eliminate enzymes and microorganisms in order to preserve the product. Heated sugary syrup can be used to preserve fruits, whereas hot brine can preserve vegetables. The food is then protected from re-contamination and light by the sealed container. A proper heat treatment and an airtight (or "hermetic") seal are essential for preservation.

There are three types of syrup: light (200g sugar per liter), medium (400-600g sugar per liter) and heavy (800g sugar per liter). Brine typically has a salt content of 15 g/1. Yeasts and moulds can be killed by pasteurization at 90-100°C for 10-20 minutes, whereas food poisoning bacteria can be killed by pasteurization at 121°C for 15-40 minutes, depending on the size of the container. Acidic fruits, on the other hand, require more severe heat sterilization to kill food poisoning bacteria. There are a number of ways that fruits can be processed and preserved until they are needed, such as in sugar syrup or sodium meta-bisulphate solution (equal to 1000 ppm). Because of the risk of food poisoning from improperly processed foods, it is not recommended that small-scale processors bottle vegetables unless they are acidified. With citric acid or vinegar as an acidity adjuster, vegetables can be pasteurized. Small-scale canning is not recommended for the following reasons: the time and temperature of canning are vital and must be closely monitored. *Clostridium botulinum*, a microorganism that can cause food poisoning and even death, can be present in under-processed cans. Cans that have been over-processed lose a great deal of their texture, colour, vitamins and flavour, making them unsuitable for sale. Proper heating temperatures are dependent on a variety of factors including food kind, can size and shape, and veggies' initial level of contamination. A trained food technologist or microbiologist is needed for this. During the canning process, food must be heated to above 100°C and the pressure inside and out of the can must be equal in order to avoid bursting cans. High-pressure steam and a 'retort' are used to accomplish this. It is unlikely that any small-scale processor will be able to afford the steam boiler and retort, which are both pricey. Additional equipment costs include pressurized air and controls required to keep can pressures constant while cooling them. For countries that do have cans, the cost is usually higher than for countries that do not. Depending on the type of product, different lacquers may be required to prevent the metal from corroding when it comes into touch with fruits or vegetables. To seal the lid onto the can, an additional tool called a "seamer" is required, as well as frequent inspections and maintenance. There are several reasons why canned food can be damaged or unsafe, but seams are one of the most common. Seamer operators must be well-versed in machine adjustment, which necessitates the purchase of a "seam micrometer" as a separate investment fee. This all adds up to a significant initial investment, well-trained and experienced staffing requirements for canning, regular upkeep of very sophisticated machinery, a steady supply of the appropriate cans, and high ongoing operational costs. A lower processing temperature is necessary because of the fruit's acidic nature, making this procedure appropriate for small-scale operations. A food technologist should always be contacted when it comes to determining how long a product should be processed and under what circumstances it should be stored once it is in a bottle.

Most of the water in fruits and vegetables is removed during drying, which increases their shelf life and convenience. Although many dried foods are delicate and must be packed in boxes to keep them from being smashed, the reduction in weight and bulk makes transportation more affordable and simpler. Dried fruits, vegetables, herbs, and spices fall into two broad categories: high-volume, lower-value crops like cereals, and low-volume, higher-value crops like those listed above. Small processors have better chances of making money in this second category.

The most popular technique of drying fruits and vegetables is air drying, and this book does not discuss more expensive methods such as freeze drying. Blanching or sulfuring/sulfiting some products can help protect their natural colour while also extending their shelf life. In order to remove some of the water before air drying, fruits are soaked in hot, concentrated sugar syrups and then crystallized, peels for marmalade and cake manufacture, and osmotically dried fruits (known as "osmasol" goods when dried in a solar dryer). Before drying, some fruits and vegetables, such as limes, can be salted. To keep food fresher for longer, the high salt concentration in this case draws out water through osmosis and also has anti-microbial qualities. Acids and flavours are produced by microorganisms that are tolerant of high salt concentrations while the product is sun dried, which limit the action of some enzymes, resulting in a loss of quality during storage. Before eating vegetables, they must be cleaned to reduce their salt content.

Dried produce must be carefully selected from a variety of fruits and vegetables. Overripe fruits, in particular, are more prone to damage and can be more difficult to dry than other types. They taste, look, and taste worse if they are under-ripe. It's critical to practice good cleanliness when preparing dried veggies since any bacteria or mould that gets on the produce before drying is likely to survive in the final product. It's not hot enough of a temperature to kill them, and they can grow again when the meal is rehydrated, resulting in food poisoning.

Blanching

Blanching kill's enzymes and preserves food's colour, flavor and texture for longer periods of time in the refrigerator or freezer. Vegetables must be dried to extend their shelf life, however this process alone does not preserve the food. Blanched vegetables are those that have been briefly heated in hot water or steam and then cooled on trays. Vegetables can be cooked in a wire basket in boiling water on a small scale by placing them in the basket.

Steamed veggies are blanched by placing them in a colander over boiling water and covering with a lid to keep the steam within. While steaming takes a little longer than water blanching, it retains more nutrients because the nutrients aren't lost to the water throughout the cooking process. Some dried fruits and vegetables can be given optional chemical treatments to preserve their colour and texture. If you want to keep the vivid green colour of green leafy vegetables like peas and okra, you can blanch them in a calcium chloride solution. If you want to maintain the texture of specific vegetables like green beans, you can blanch them in sodium bicarbonate. Pharmacies in major cities generally carry both chemicals.

Sulphuring and Sulphiting

On average, 350-400g of sulphur per 100 kg of fruit is burned for 1-3 hours for most fruits. Sulphur dioxide can be used to avoid browning in foods such as apples, apricots and coconuts, however it should not be used on red fruits because it can lighten their colour. In a sulphuring cabinet, chopped or shredded fruits are exposed to burning sulphur while being sulphured (using sulphur dioxide gas). To determine how much and how long the fruit is exposed to sulphur, it is necessary to take into account the type of fruit, its moisture content and the legal or commercial limits in some countries on residual sulphur dioxide levels in the finished product. The Bureau of Standards should be contacted to verify this information. Sulphite-treated fruits are becoming less popular with consumers in some industrialized countries, so the Export Development Board or import agents should be informed before exporting the product.

Instead of using the gas meta-bisulphite is dissolved in water in sulphiting, either by adding one or more of them to the blanching water or more frequently, by soaking the food for 5-10 minutes in sulphite dips. When sodium meta-bisulphite is dissolved in water, almost two-thirds of its weight is converted into sulphur dioxide.

For example, 1.5g of sulphur dioxide is dissolved in a liter of water to produce 1g of sulphur dioxide per litre in a 0.001% solution, which is equivalent to 1000 parts per million. As long as the sulphiting solution is at this concentration, it can be used to store fruits in an intermediate state for several months at a time.

Syrup pre-treatment

Because it removes half of the water in fruit, it's a low-cost option to boost drier production or part-process fruits for interim storage, allowing producers to keep up with demand throughout the entire year. As a result, the dried food has better colour retention and is sweeter and blander tasting thanks to this technique. Because fruits lose their acidity when they're dried and wrapped properly, mould can form on food that isn't properly dried or packaged. The fruit is initially cooked in 20 percent syrup before being steeped overnight in the more complicated procedure. Once the fruit has been filtered from the syrup, it is then moved to 40 percent and 60 percent syrups in sequence, with each transfer requiring a 10 minute boil. It is necessary to dilute the syrup by half before serving it. New 60 percent syrup is produced every day by using the diluted syrup (10 percent) in other goods. Reusing sugar syrups and creating a softer texture are two of the benefits of this procedure.

Types of dryers

The higher value of dried fruit and vegetable products, compared for example to cereals crops, may justify the higher capital investment in a fuel-fired dryer or electric dryer and crops, may justify the higher capital investment in a fuel-fired dryer or electric dryer and the extra operating costs for the fuel or electricity. These types of dryer allow higher drying rates and greater control over drying conditions than do solar or sun drying and they can therefore result in a higher product quality. However it is necesary to make a careful assessment of the expected increase in income from better quality products compared to the additional expense, to make sure that this type of dryer is cost-effective.

Sun drying is only possible in areas where, in an average year, the weather allows foods to be fully dried immediately after harvest.

The main advantages of sun drying are the low capital and operating costs and the fact that little expertise is required. The main problems with this method are as follows:

» Birds, rats, or insects may cause contamination, theft, or damage mould can grow more easily if there is no protection from rain or dew on the product, which can lead to higher final moisture content if the drying process is too sluggish or intermittent.

» The low and varying product quality caused by over- or under-drying of large expanses of land required for the shallow layers of food because the crop must be turned, shifted if it rains, and animals must be kept away, it is laborious.

Some fruits and green vegetables lose their quality (colour and vitamin content) when exposed to direct sunshine. By lowering the size of the pieces and drying them on raised platforms, wrapped in fabric or netting, the quality of sun-dried foods can be improved.

In spite of extensive research by academics from around the world, solar drying has yet to see widespread commercial application. Main uses thus far include desiccated coconut in Bangladesh; herbal tea from Guatemala; and dried fruits for export out of Uganda.

Packaging

Packaged dried foods may not be essential in dry climates because they will not absorb moisture from the atmosphere. However, in a humid environment, dried foods are more likely to get damp and moldy. In addition to the humidity of the air at which a food does not grow or shed weight, the type of food also influences the stability of dried foods. It is possible to categories meals based on their ability to absorb moisture from the surrounding environment. Hygroscopic items, like non-salt and pepper, are able to quickly absorb water, whereas fruit and vegetable products are also hygroscopic. Various fruits and vegetables necessitate different packing requirements because of this. Equilibrium Moisture Content (EMC) is the moisture content at which a food is stable.

In most cases, plastic film is used to protect dried fruits and vegetables. The following factors must be taken into account when deciding on the best sort of packing material:

» The air's temperature and humidity, as well as the product's ability to absorb moisture from the surrounding atmosphere.

» During the product's specified shelf life, reactions within the product are produced by air or sunlight.

» Aspects of the marketing strategy.

» The cost and availability of different packing materials.

Polythene film is inexpensive and widely accessible; however it can only be used to store dry fruits and vegetables for a limited period before they become soft and mouldy due to its susceptibility to moisture. Yet, polypropylene is more expensive than nylon and has a shorter shelf life; however, polypropylene may not be available in all regions. Dried foods are better protected by laminated polythene and aluminium foil coverings, although they are more expensive and difficult to get in developing nations.

Most dry foods require a solid box or carton in order to prevent crushing and to exclude light, which causes colour loss and the formation of off-flavors while in storage, as well as to keep the food fresh.

Chutneys, Pickles and Salted Vegetables

Chutneys

Many different fruits and vegetables are combined with sugar, spices, and sometimes vinegar to create chutneys. Any sour fruit that is edible can be used as a base for chutney in order to counteract the sweetness of the sugar in the dish. Sugar acts as a preservative, thus vinegar isn't always necessary; this can depend on the fruit's natural acidity and ripeness. Boiling most items results in caramelized syrup, which modifies the flavour, colour and thickness of the product, as well as pasteurizing it, which increases the sugar and acid preservative effect. Products that ferment naturally are protected by the acids produced by a variety of microorganisms. Spices, in addition to imparting flavour, can serve as preservatives depending on the varieties used. In order to preserve the chutney after the jar has been opened, it uses natural acids from the fruit and high sugar content. Preventing mould formation necessitates the use of a Preservation Index to calculate the appropriate amount of sugar and acid to be applied. A refractometer can be used to ensure that the final sugar content of the syrup is between 68 and 70 percent if sugar is the primary ingredient or the product is boiled. Prior to heating if a dark product is to be produced, or at the end of the boiling process for a light coloured product, the sugar is added.

Combinations of acid and sugar are quantified by the Preservation Index (sugar is measured as "total solids"). Food deterioration and food illness microorganisms can be detected using this test. Here's how to figure it out:

When basic laboratory equipment is not available or the maker is unsure of how to perform the calculation, it is recommended to send a sample of the product to a Bureau of Standards, University Food Science Department or food testing laboratory for analysis.

Pickles

Lactic acid bacteria, which can grow in low salt concentrations, ferment vegetables like cucumber, cabbage, olive, and onion. Lactic acid is produced as a byproduct of the bacterium's fermentation of sugars in the food, and this inhibits the growth of pathogenic bacteria, moulds, and other spoilage organisms. In order to manage the kind and rate of fermentation, the amount of additional salt must be applied. Lactic acid is produced by a natural sequence of different species of bacteria when, for example, 2-5 percent salt is employed. 'Salt stock' pickles, which are preserved by the salt rather than the fermentation process, can be made with higher salt concentrations (up to 16 percent). As an intermediate storage method, fruits and vegetables can be stored this manner to extend the year's harvest by several months. Sugar can be used to speed up fermentation or sweeten the final product. Alternatively, vinegar (acetic acid), salt, and sugar can be used to preserve vegetables, resulting in a variety of pickled foods. Their flavour and texture are distinct because they haven't been fermented. Pasteurization is common. Single fruits or a combination of fruits and vegetables can be used to make sweet pickles. In order to keep them fresh, lactic or acetic acid, sugar, and other spices are used in combination.

Single fruits or fruit and vegetable mixes are used to make sweet pickles. The combination of lactic or acetic acid, sugar, and, in some circumstances, other spices preserves them.

Salted vegetables

Salted veggies are created by layering chopped or shredded vegetables such as cabbage with salt in a covered drum and cooking them for a long period of time. One preservation function is to extract water from the veggies through osmosis, and the other is to work directly against microorganisms by creating concentrated brine in the drum's bottom. Washing the food reduces the salt content before consumption. No pasteurization is required for pickles that have a high Preservation Index. Pasteurization or heating the sugar/salt/vinegar mixture and then adding it to the veggies and filling the jars while the result is still hot are both options for preventing spoiling. When the hot substance cools, it creates a partial vacuum inside the jar, which aids in preservation.

If the pickle's shelf life is projected to be less than that of the glass jar, it can be packaged in polythene pouches and sealed with an electric heat sealer. There are several ways to prevent product leakage, which can destroy paper labels and make the container unappealing, but a double pouch can be utilized to avoid this problem.

Pectin and Papain

Pectin can be isolated and used in food processing to create the distinctive gel in jams and marmalade because it is found in practically all fruits and vegetables. Citrus fruits like lime, lemon, orange, and passion fruit, as well as apple pulp, have the highest concentrations of pectin. Both powdered and liquid forms of pectin are commercially accessible. In a cool, dry area, it will lose just around 2% of its gelling strength per year if it is stored properly. Pectin can be divided into two categories:

1. Pectins with a high methoxyl (HM) content that gel in a pH range of 2.0-3.5 in high solid jam (above 55 percent solids).

2. The calcium salts in low methoxyl (LM) pectins are used instead of sugar or acid to generate a gel.

In a pH range of 2.5-6.5 the solids range from 10% to 80% forming a gel. Spreads and gelling agents in milk products are the most common use for these ingredients. It's important to note that there are numerous variations in pectin within each category, such as "quick set" and "slow set" varieties, so be sure to specify which type you need when placing an order. It is difficult to extract the raw pectin in their products without the use of solvents using only water. Fruits like melons, which have low pectin content, can benefit from this. Pectin can be used to manufacture set preserves, such as jams, jellies, etc., if the pH is between 3.0-3.5 and the solids content is at least 68 percent. Pectin powder is added to fruit pulp at a rate of 3-6g per Kg of finished product, but it must first be mixed with around five times the weight of sugar in order to avoid lumps developing when it is added to the pulp and juice.

Papain

Skins of unripe papaya fruits contain papain, an enzyme. Industrialized countries use it extensively for meat tenderization and brewing because it breaks down proteins. In the last two decades, advances in biotechnology have produced synthetic papain that is cheaper than natural enzymes, and as a result, the market for natural papain has declined. Natural products, however, have just re-emerged as a hot commodity and may now command a higher price than ever before. Papaya trees are frequently grown in orchards rather than collected from a wide variety of scattered trees, which increases the cost of gathering the fruit. Because refining equipment is out of reach for most small-scale enterprises, crude papain is typically generated. Making shallow cuts in the skin of unripe fruits while they are still on the tree is required for papain extraction. Scraping the fruit on a daily basis for several weeks causes the skin to exude a white, sticky liquid. It is then sun-dried till the 'latex' becomes hard and brittle. Packaged crude papain is exported to refiners for processing. In order to protect the operator's skin from papain, the crude latex should be handled using gloves.

Sauces

Salted, sugared, spiced, and vinegar pulped fruits and vegetables are used to make thick, viscous sauces. Vinegar is used as a preservative since it prevents the growth of spoilage and food-poisoning microorganisms, which is why they are pasteurized. In addition to salt and sugar, the right Preservation Index ensures that the product does not decay after opening and that it can be used a little at a time without compromising its preservation effect. Some may contain sodium benzoate as a preservative, but this isn't necessary if the Preservation Index is enough. Although sauces can be prepared from nearly any combination of fruit or vegetables, the market is dominated by tomato sauce, chilli sauce, and to a lesser extent, mixed fruit sauces like 'Worcester' sauce, which includes apples and dates. The Preservation Index can be obtained depending on the scale of production. Tomato sauce, chilli sauce and to a lesser extent, mixed fruit sauces like 'Worcester' sauce, which incorporates apples and dates in addition to tomatoes, are the most popular sauces in many countries. However, other fruit and vegetable combinations can be used to create sauces. Processing seeds and skins by hand or with sophisticated pulper-finisher equipment depends on the volume of the production. Sauces can also be created using basic open boiling pots, provided that care is given to heat gently with continual stirring to avoid localized scorching of the product, particularly at the conclusion of heating. Steam-heated stainless steel 'double jacketed' pans are used for larger-scale processing.

Juices

Juices, nectars, and carbonated drinks, which are intended to quench thirst, make up the bulk of the beverage market, while wines and spirits, which are reserved for special occasions or consumption by religious or cultural custom, make up the remainder. Beverage businesses face intense competition from large-scale producers, and as a result, significant sums of money are spent on advertising, packaging, and advanced delivery networks. As a result, beverage manufacturing is one of the most challenging for small-scale businesses to develop and succeed. However, greater juice consumption is on the rise in some emerging country cities, and this market may grow in the coming years.

When it comes to drinks, there are those that are meant to be consumed straight from the bottle and those that are meant to be diluted before being consumed. Preservatives, such as sodium benzoate, may be necessary in the second group if the first group is to have a lengthy shelf life after opening. Depending on the storage conditions, both types of unopened bottles have a shelf life of 3 to 9 months. If you're going to use citric acid to raise the acidity level of melon juice, you're going to have to use a lot of it, thus you're going to need a lot of citric acid. Pasteurization and the juice's natural acidity help preserve it. Pulped fruit or juice can be used to make a broad variety of beverages, and production can be carried out over a longer period of time by processing a succession of fruits or by part-processing pulps and preserving them in 1000-2000 ppm sodium meta-bisulphite solution.

Hand-pulping soft fruits like berries or passion fruit, while more fibrous fruits like pineapple and mango necessitate the use of a machine for pulping. Where electricity isn't available, manual pulpers are an option, but powered pulpers and blenders are also readily accessible. Higher output rates require pulper finishers to brush fruit through a sieve in order to separate the pulp from the seeds, peels, and so on. Fruit can be pressed or milled to obtain juice, or the juice can be extracted by heating the fruit. Reaming the fruit extracts the juice, and again, this may be done using relatively inexpensive equipment.

Pasteurization

If a beverage is to be stored for longer than a few days, it must be pasteurized. In general, pasteurization takes between 10 and 20 minutes at 80 to 90°C depending on the type of substance and the size of the bottle. A big pan of simmering water with the water level just above the bottle's shoulder can be used for both hot filling and cold filling, and both methods can be used for pasteurization. The bottles are unlikely to rupture because to the low pasteurization temperature of less than 100°C. As with other glass containers, there should be no more than a 20°C temperature difference between the product or hot water and the glass container. Using a water cooler, full containers can be cooled more quickly. To avoid 'heat shock' to the containers, a counter-current flow of hot water and cool water is used in a trough. All bottles must be thoroughly cleaned to avoid contamination by dust, insects, etc., using either hand-held bottle brushes or automated brush cleaners. Bottles that are going to be re-used need to be properly cleaned with detergent to eliminate any leftover material and then sterilized for at least 10 minutes in a boiling water bath.

Filling

Fillers can be created by attaching one or more taps to the bottom of a stainless steel or food-grade plastic bucket. Hand filling is typically too slow for the needed throughput. In more advanced fillers, the volume of liquid injected into each bottle is measured and controlled. As a result of their reduced cost, waxed cartons have become a popular alternative to bottles, saving both time and money in the collection and washing of unused bottles. There are certain countries where small-scale producers can't afford the machinery needed to make and seal cartons; instead they're turning to cheaper alternatives like plastic containers with sealed foil tops.

Squashes, Cordials and Syrups

Squashes and cordials

These are watered-down beverages that can be enjoyed in small doses. To prevent spoiling after opening, the container of these products must be resalable and preservatives, such as sodium benzoate, may be included. There must be at least 30% fruit juice and sugar syrup in a squash. Simply put, cordials are squashes that are crystal clear. Some manufacturers utilize food colours, however this isn't necessary for the majority of their products. The composition of squashes is regulated in a large number of nations.

Syrups

Syrups are made by boiling filtered liquids until the sugar level reaches 50% to 70%. Sugar or honey can be substituted for syrup because of its high solids content and preservation properties. Grape syrup is the most prevalent sort of syrup, while syrups prepared from other fruits are also common.

Preserves

Jams, Jellies and Marmalades

The pulp or juice of a single fruit or a variety of fruits is used to make jam, a solid gel. In some nations, the composition is regulated by law. The first-named fruit should make up at least half of the overall fruit content in mixed fruit jams. After opening the jar, the sugar concentration is typically 68-72 percent, which prevents mould from forming. Jellies are delicious. Clear jams made with filtered juice rather than fruit pulp. Citrus juices are used primarily in the production of marmalades, which contain small slivers of citrus peel suspended in the gel. As well as citrus fruits, ginger can be utilized on its own or in combination. Citrus fruit should account for at least 20% of the fruit, and the sugar concentration should be comparable to jam. In order to achieve the desired gel structure, the precise ratio of acid, sugar, and pectin must be used, and rapid boiling is required to evaporate water quickly and concentrate the mixture before it darkens and loses its gel-forming potential.

Pastas and purees are both types of preparations

Any type of fruit or vegetable can be used to make pastes and purees. Tomato and garlic are two of the most prevalent varieties. Concentrated pulp can be manufactured on a small scale by carefully evaporating water and stirring constantly to avoid darkening or localized burning. Typically, the paste has a solids content of 36% or less (by weight). Pasteurization in bottles or cans is required for a longer shelf life due to the product's high solids content and natural acidity. Sugar, salt, and vinegar are common preservatives included in various dishes.

Fruit-based cheeses

They are fruit pulps that are heated to a sugar level of 75 to 85 percent, and then pressed into cheese. In contrast to jams and marmalades, they do not have the gel structure that sets as a solid block. Alternatively, they can be cut into bars or cubes and eaten as is, or they can be utilized in confectionary or baked goods as small bits.

Preparation of a batch

The Pearson Square is a simple method for determining how much sugar and juice to use in squashes and jams. Squash with 15% sugar content, created from orange juice with 10% sugar and 60% syrup, can be made using the Pearson Square in the following way: Write the juice and syrup concentrations on the left and the product concentration on the middle of a square. To determine the proportions to be combined, take the smaller amount and divide it by the bigger amount diagonally. Similar Pearson Squares can be used to determine how much of each component should be used in a mixture.

Boiling

A stainless steel pan is used to bring water to a boil. There is a chance that fruit acids will react with the pan and produce off flavors' if other materials are utilized. The consistent and rapid heating provided by a steam jacketed pan is preferable at greater production rates. There are two stages to making jam if whole fruit is used. First, the fruit is heated gently to soften the flesh and extract pectin; then the combination is boiled rapidly until the sugar concentration reaches 68% to 72%. This shift in heat output necessitates a heat source that is large enough and easily manipulated. A sugar thermometer reading of 104-105°C.

When preparing items like tomato paste, hanging the pulp in a sterilized cotton sack for an hour is an alternative approach to boiling. During this time, the pulp loses half its weight due to the loss of watery fluid. After that, a salt solution of 2.5 percent is added to the concentrate, and the weight is reduced to one-third of what it was originally. A variety of options are available for packaging, pasteurization, and concentration. This process is said to use less fuel than boiling concentration while yet producing a natural-tasting result. New or re-used glass jars should be filled with preserves and then sealed with a new lid. However, plastic containers are now being used in some nations; however the seals on these containers are sometimes poor, resulting in product leakage and insect infestation issues, as well as a reduced shelf life. Filling should be at or above 85°C. Steam condenses on the lid's interior and waterfalls over the preserve's surface if the temperature is too high. This leads in the growth of mould because the sugar on the surface is diluted. Because of the thickening that occurs if the temperature is too low, it is difficult to put the preserve into containers. Small hand-operated or semi automated piston fillers can be used for filling jugs and funnels at greater production rates. As the product cools, the jars should be filled to about 9/10th of their capacity to help create a partial vacuum. A "headspace gauge" can be used to verify this. During the cooling process, the jars must be kept upright until the gel has formed.

WINES, VINEGARS AND SPIRITS

Wines are made by fermenting fruit juice/pulp and additional sugar into alcohol and carbon dioxide using a variety of the yeast *Saccharomyces cerevisiae*, referred to as"wine yeast". In order to consistently produce a high-quality product, manufacturers must choose one type and stick with it. The high levels of alcohol and natural acidity in wine help to preserve it. Some of the most common fruits for making wine in many poor nations include grapes and pineapple, papayas, grapefruit (passion fruit), bananas, and melon. 'Fortified' wines, such as sherry, ginger wine, etc., often have an alcohol concentration of 15-20 percent. In order to avoid wine spoilage caused by other microorganisms, it is critical that fermentation vessels be thoroughly cleaned and sedimentation or filtration is performed to ensure a crystal-clear end product. In most nations, a special license is required to sell alcohol. Acetic acid bacteria (*Acetobacter species*) convert wine's alcohol to acetic acid during a second fermentation to generate vinegar. This second fermentation necessitates as much air contact as possible, in contrast to winemaking, where an air lock keeps out oxygen. Letting the wine to flow through an open framework, or "generator," which allows the air to freely circulate, is the traditional method of allowing the wine to age. An airtight container is required to keep acetic acid from being evaporated from the wood acid, which might preserve a product for months or even years. In more modern vinegar fermenters, air is pumped into the wine, and they are so expensive that most small producers cannot afford them. In order to make spirit drinks, distillation is employed since alcohol has a lower boiling point than water. Clean drums with safety valves and exhaust pipes serve as stills for distillation processes. The drum is filled with wine or other alcoholic beverage and heated. The distillate is recovered after the alcohol vapour passes through cooled air or cool water. A word of caution: distilling without a government permit is illegal in several nations all together.

Distillation can be done in a plethora of ways that date back centuries. East Africans use an oil drum filled with fruit wine or beer and a shallow horizontal angle over a fire. The drum's aperture is sealed with a bamboo pipe about 10 cm in diameter and a pulp formed from the inner layers of a banana stump. They fit vertically down from the bigger pipe and seal in the same manner as the five smaller bamboo pipes. Each pipe is connected to a 20-liter baked clay pot that is buried in a slow-moving pit of diverted water from a stream. Wood branches are folded over the pots and inserted into the soil at either end to keep them submerged in the water. Each pot is sampled at regular intervals to determine its alcohol level and the rate of heating is adjusted accordingly by adding or withdrawing firewood. Distillation is controlled the desired end determines the final point of distillation as the strength decreases with each subsequent distillation. To regulate the distillation process, fuel is added or removed and samples of each pot's alcohol strength are taken at regular intervals. Distillation continues to weaken the product's potency until it reaches a targeted level of 35 to 40 percent alcohol. Thirty-four liters of spirit are collected from the pots and placed in clean containers before being sealed with a stopper made of banana or papyrus leaves. By virtue of its high alcohol content, the spirit is well-protected and can be stored for several years.

Stainless steel and copper are the materials of choice for modern stills, which have thermostatically regulated heaters. Small copper stills for manufacturing essential oils have been used to make small amounts of alcohol, but they are too expensive for most small-scale manufacturers. Adding sugar to fruit juice to elevate the level to 20° Brix and then inoculating it with approximately 3% yeast for about ten days in a food-grade plastic or glass fermentation vessel is common practice. To prevent the batch from being infected by germs and mould, it is essential to keep the container sealed. Wine is 'racked' (filtered) by passing it through a muslin or nylon straining cloth into narrow-necked fermentation vessels that are plugged with cotton wool or fitted with an air lock after 10 days of fermentation. Depending on the temperature and the strain of yeast employed, the fermentation might last anywhere from three weeks to three months. When no more bubbles can be seen rising to the surface, the fermentation is complete. Next, it's put into sterilized bottles with sterile cork stoppers or roll on pilfer-proof (ROPP) screw caps and left to clear and develop. Natural gums in the fruit of some fruit wines, such as pineapple, make them difficult to remove. The gums can be precipitated out of the juice before fermentation by boiling the juice. Waste fruit from other operations (such as drying or bottling) might provide a difficulty when employed in winemaking. Only the fermented liquor should be used for distillation at all times. Because of the possibility of methanol contamination, any suspected liquor or other materials should not be distilled. Blindness and death can result from excessive consumption of this sort of alcohol. A sanitary and easy-to-maintain structure is essential for all food processing processes, including those involving fruits and vegetables. Insects and animals, as well as microorganisms, are the two primary sources of contamination. The presence of food waste attracts insects and other creatures to food structures after production has ended. Equipment, tables, and floors that have not been adequately cleaned can harbor microorganisms. In order for microorganisms to proliferate, they need water; hence wet processing is more susceptible to contamination than dry processing. Strict hygiene should also be enforced during drying operations since some microorganisms can generate inert spores that can survive in dry circumstances and then multiply when they come into touch with water or food. Dry processing presents an extra danger of contamination by dust, which can both degrade food and harbor microorganisms. Entrepreneurs planning to develop a new processing facility or renovate an existing one should consider the following factors before beginning construction.

The Site

When deciding on a site for a food building, the following factors should be taken into account:

» Supply and potential markets in terms of geography.

» Staff accessibility (public transport, distance down an access road).

» Quality of road access (all year, dry season only, potholes that may cause damage to products, especially when glass containers are used).

» Swampland that would be a source of odours and insects in the immediate area.

» Upstream of the processing facility, any available ground for waste disposal away from the building has the potential to be contaminated.

» Supply of electricity.

» Cleared the ground to reduce insect and bird concerns (preferably planted with short grass, which acts as a dust trap for airborne dust).

Design and Construction of the Building

Buildings must be large enough to accommodate all production processes, as well as storage space for raw materials, packaging materials (and finished items), and finished goods. However, the investment should be proportionate to the size and predicted profitability of the firm in order to minimize start-up capital, the size of any loans taken out, and depreciation and maintenance costs.

Roof and ceiling

Overhanging roofs in tropical areas protect the walls and the building from direct sunshine. With heating, this is especially crucial to ensure a more pleasant working environment. Compared to galvanized iron sheets, fibre-cement tiles provide better heat insulation. Fresh air is drawn into the processing area by the use of roof vents at high levels. To keep insects, rodents, and birds out of the space, the vents must be screened with mesh. The entrepreneur may want to explore installing electric fans or extractors if heat is a major issue, although this obviously increases capital and operational costs. Processing and storage rooms should be free of rafters or roof beams. Toxic dust might build and fall off in lumps, contaminating the products. Insects, too, can fall off them and end up in the final product. There are also concerns of contamination by hair, feathers, or excreta from rats and birds because of their access to these areas. This means that panelled ceilings are important in any processing or storage area, and that they must be installed with care so that no holes can be found. Additionally, openings in the roof structure or where the roof connects to the walls should be sealed to keep flying insects, birds, and rodents out of the processing area.

Walls

All inside walls must be rendered or plastered with a high-quality plaster to prevent dust from accumulating in the processing area. To avoid the accumulation of dirt and insects, engage an experienced plasterer to smooth out any remaining cracks or ledges in the surface finish. A specific focus should be placed on making it easy to clean the bottom portion of the walls, up to a height of at least 1.08 meters (four feet) above the floor, where washing equipment and product splashes are most likely to occur. Emulsion should be used to paint the higher parts of the walls. To make them easier to clean, the bottom portions of walls should be either coated white with waterproof gloss paint or tiled with glazed tiles. The cost of tiling a process room can be reduced by tiling only certain sections, such as beneath sinks or machinery. The Ministry of Health or other suitable body should be consulted if there is a legal requirement for specific interior finishes in your country.

A sloped window sill helps keep dust from gathering and prevents operators from putting cloths or other objects there, which in turn might attract insects. Instead of using expensive electric lighting, offices with windows benefit from free, healthy natural light. Workers in hotter climes, on the other hand, are more likely to leave windows open to let in more outside air. As a result, flying insects have easy access to the goods and can cause contamination. In order to allow windows to be left open, mosquito mesh should be installed. Normally, doors should be locked, but if they are often used, they are more likely to be left open, leading to the same problem of animals and insects entering the plant. Insects and some animals may be deterred by a thin metal chain or strip of material that is suspended vertically from the door lint l. Instead of mesh door screens, you may want to utilize them. It is important to ensure that all storeroom doors are closed to avoid insects and rodents from spoiling stock or ingredients by keeping the doors shut.

Good quality concrete, smooth finishing and a lack of cracks are all must-haves when it comes to the floors of processing and storage areas. Proprietary floor paints and vinyl-based coatings can be purchased in some poor nations, although they are typically highly expensive. Using the red wax floor polishes that are widely accessible in homes is generally not sufficient, as these wear away quickly and potentially contaminate either items or packages. Acidic fruit juice spills deteriorate concrete over time because they react with it. As a result, spills should be cleaned up right once, and the state of the floor should be checked on a regular basis. When the floor and the walls meet in a corner, dirt tends to collect in such areas. The floor should consequently be bent up to meet the wall during construction. The right angle can be filled by placing concrete fillets in the corners of an existing floor, but care must be taken to prevent the creation of new gaps that would harbor dirt and insects. At an angle of approximately one-eighth of a degree, the floor should slope toward a central drainage channel. The floor can be completely cleaned and drained at the conclusion of each workday. In order to avoid the danger of equipment and food contamination due to stagnant water, proper drainage is essential. The drain should be equipped with a steel grating that can be readily removed so that the drain is cleaned. If wire mesh is not installed over the drain entrance, rodents and crawling insects could gain access to the building. This, too, should be able to be removed for cleaning purposes.

Lighting and power

Lighting in the general room should be kept to a minimum as much as practicable. The cost of running fluorescent tubes over incandescent bulbs is significantly lower when additional lighting is required. While incandescent lights should be used for machines with fast-moving exposed parts, tubes should be avoided. There are apparent risks to workers even though the parts should be guarded, as long as the speed of a spinning machine matches the frequency of mains electricity, which powers fluorescent tubes.

When cleaning the floor or equipment, make sure that all electrical outlets are elevated above the ground so that water cannot leak into them. Ideally, water-resistant sockets should be employed. Each outlet should only be used for one thing at a time to avoid overloading a circuit and igniting a fire. Additional power points should be installed, even if they are more expensive, if there aren't enough available to meet the demands of a process. Ideally, the mains supply should have an earth leakage trip switch installed, and all connectors should be equipped with fuses appropriate to the power rating of the equipment. Using three-phase power for larger machinery or heavy loads from electric heating requires that the wiring be built by a trained electrician to ensure that the supply is evenly distributed throughout the three phases. Water is essential in nearly all fruit and vegetable processing, both as a component of products and for cleaning. Therefore, taps around the processing area should have an appropriate supply of drinkable water. It is vital for a business to make plans to ensure a daily supply of high-quality water in many nations where the mains supply is inconsistent or contaminated. Two high-level, covered storage tanks can be installed either in the roof area or on pillars outside of the structure to accomplish this goal. When mains water is available, both tanks can be filled at the same time, allowing the sediment in one tank to settle while the water in the other is being used. Sedimentation takes a long time, thus each tank's capacity should be enough for a single day's worth of production. Drain valves should be installed above the slope and at the lowest part of the tanks, which is why they should have a sloping base. When the tank is nearly empty, the lower valve is opened to flush out any sediment that has accumulated. Water is drawn from the upper valve during use.

Any water used in a product should be thoroughly treated to eliminate any sediment and if necessary, sterilized before being used in the final product. If the product is not heated once water is added as an ingredient, this is very crucial.

It is possible to purify water on a modest scale using four methods: filtration; heating; ultraviolet light and chemical sterilants. At a small scale, other water purification procedures are too expensive. Domestic water filter filtration is sluggish, but once you've invested in it, it's reasonably inexpensive. In some countries, larger industrial filters can be found. Heating water to a rolling boil and holding it there for 10 to 15 minutes is a straightforward and low-cost process, but it is costly and time-consuming to perform on a regular basis because of the expenses of energy. Boiling water does not remove sediment, therefore it may need to be filtered or allowed to stand for a while to remove it. Finally, hypochlorite chemical sterilization is quick, affordable, and effective against a wide spectrum of microorganisms. The amount of chlorine in cleaning water should be 200 parts per million, whereas the amount of chlorine in ingredient water should be 0.5 parts per million in order to prevent the product from tasting like chlorine. Adding 1 liter of bleach to 250 liters of water yields a chlorine concentration of 200 parts per million, while 2.5 ml of bleach to 250 liters of water yields a chlorine concentration of 0.5 parts per million. There are several drawbacks to the use of chlorine, including the corrosion of aluminium equipment and the tampering of foods with bleach. The concentration of chlorine can be detected in water using a chemical dye that changes colour when it reacts with chlorine. With the use of a "comparator," the hue and saturation of the image are compared to those of standard colours on glass discs.

To limit the danger of product contamination and to repel pests, rodents, and birds, sanitation is vital. Instead of piling up trash on the floor, make sure it goes in trash cans. Instead of allowing waste to gather throughout the day, processes should be equipped with a waste management system that removes waste from the facility as it is created. Leaving wastes in a processing room overnight is a bad idea. To limit the danger of product contamination and to repel pests, rodents, and birds, sanitation is vital. Instead of piling up trash on the floor, make sure it goes in trash cans. Rather than allowing wastes to accumulate throughout the day, processes should have a management system in place to remove wastes from the facility as they are produced. Leaving wastes in a processing room overnight is a bad idea.

Layout of equipment and facilities

There should be a physical separation between different stages of a process wherever possible. So that completed products aren't contaminated by arriving, frequently filthy raw materials and sections of the space where extra cleanliness attention is needed are easily identified. Preventing contamination from activities such as bottle washing, when inevitable breakages result in glass splinters, is especially critical. As a bonus, accidents and operator collisions are less likely as a result of the extra space. Non-perishable ingredients and packaging materials should be kept apart from perishable raw materials. A peaceful working environment for book keeping is made possible by having a separate office where records may be filed and maintained. Either a separate building or two doors separating the toilets from the processing area are required. Towels and soap should be provided for all staff to wash their hands. Fruit and vegetable processing does not necessitate laboratory facilities, but a separate table for quality assurance checks or check-weigh packages of finished product could be put in the office or in a different part of the processing room.

The size of the equipment used should correspond to the desired production scale. Managers should create and enforce regular cleaning and maintenance programmes. Processing raw materials like fruit and vegetables necessitates the use of basic equipment including buckets, tables, knives, and scales. A +/- 0.lg scale for weighing small quantities and a +/- 50g scale for weighing greater quantities of raw materials are both ideal. The price of scales varies greatly from country to country, however the cost of compact, electronic scales is decreasing. If you don't want to spend money on a scale, you may simply calibrate scoops or other measuring devices so that they hold the exact amount of material when they're full. Slower than weighing, scoops can be used in place of weighing, but they are less accurate and require more training for operators. Food contact equipment should be built from food grade plastic, aluminium or stainless steel due to fruits' acidic nature. To avoid off-flavors or colour changes in the product, mild steel, brass, and copper should not be utilized. In general, stainless steel is used only for cutting blades, boiling pots, etc. due of its expensive cost. Despite the fact that wooden tables are more expensive than metal ones in most nations, they are more difficult to clean. A 'melamine' surface or a thick plastic sheet should be used to cover wood in order to make cleaning easier.

Instead, they can be placed in a steam chamber on a wire mesh. They are often sulphured or sulphited to preserve the colour of fruits. These can be simply manufactured from readily accessible materials in the area, such as food-grade plastic tanks for the sulphite dips and simple wooden boxes for the sulfuring cabinets. There are a variety of equipment that may be used to prepare fruits and vegetables at greater scales. Cleaners, de-stoners, peelers, cutters, and slicing or dicing devices are all included in this list. In order to boil the syrup, food-grade plastic tanks and aluminium pans are used to soak the crystallizing fruits in syrup first. Syrup is concentrated over a period of three or four days using a series of tanks, which allows for better sugar utilization compared to single-stage incubation. A muslin cloth filter should be used to remove dust and other pollutants from sugar in most countries.

The faster drying minimizes the chance of spoilage, enhances the product's quality, and reduces the drying area that is required. However, care must be taken when drying fruits in order to avoid hardening of the case and subsequent mould growth as a result of overly quick drying. Using a solar dryer will keep your food free of dust, insects, birds, and other creatures that would ruin it otherwise. They're easy to build and don't require any fuel, thanks to readily available local resources. As a result, they may be of use.

» Where fuel or electricity are expensive, erratic run available.

» Where land for sun drying is in short supply or expensive.

» Where sunshine is plentiful but the air humidity is high and as a means of heating air for artificial dryers to reduce fuel costs.

As fuel prices rise or our dependency on foreign fuels decreases, this last application is likely to become more important. The quality of sun dried foods is acceptable to local consumers and the additional expenses of solar drying are not repaid by increased food value, so it is unlikely that solar drying will be useful.

Fuel-fired dryers must be used when solar dryers fail to provide adequate control over drying conditions. When it comes to the many types of furnaces available for purchase, there is a wide range of options to choose from. Additionally, fuel-fired dryers are more difficult to construct and maintain, resulting in greater operating and capital expenses. They also necessitate specialized labour to operate and maintain.

The cabinet dryer, which may be used to dry herbs, herbal teas, and spices, as well as fruits and vegetables is a popular drying method in many developing nations. Generally speaking, a drying space of 1 square meter is required for 2-6 Kg of raw materials; depending on the type of food (6 kg of chopped fruits require 1 square meter, while shredded cabbage can only be placed at around 1 square meter).

Cabinet dryers, who can be used to dry herbs, spices, herbal teas, and other herbal products as well as fruits and vegetables, are popular in developing nations. While a product like shredded cabbage is less dense and therefore can only be piled at a density of 2 kg/m, dried fruits and vegetables require a drying space of lrn2 and can be dried in an area as small as 2 kg/m.

Boiled, Concentrated and Pasteurized Products

Juices, squashes, sauces, pickles, and chutneys are all part of this collection. To make juice or pulp, there are a variety of ways to extract the juice or pulp from fruits and vegetables. It is possible to juice soft fruits and vegetables such as berries using a juicer attachment to a food processor, as well as using a fruit press. To remove the bitter pith or skin from citrus fruits, they are typically reamed. Using a liquidizer or a large-scale pulper-finisher, fruits such as passion fruit and tomato are peeled and then pulped to remove the skins and seeds. Some chopped soft fruits, such as melon and pawpaw, can be 'dissolved' in steamers, such as those used for blanching. A fine muslin cloth or stainless steel juice strainers must be used if you want a clear juice.

In order to pasteurize or condense most goods, heat is required. A stainless steel pot for boiling water is required in all circumstances. In many nations, local fabrication is challenging due to the lack of welding skills and facilities for stainless steel. To be sure, there are other ways to go, but it's essential that a product be made of the best quality possible. There are some situations when it is possible to save money by heating syrup to boiling in a large aluminium pan and then mixing it with juice in a smaller stainless steel pan to obtain the requisite pasteurization conditions, such as squash manufacturing, for example.

A simple stainless steel pan can be put directly over the heat source at lesser scales of operation, while a larger pan can be used for larger operations. An indirect heated pan, or "double jacketed" pan, can be utilized for large-scale production. Since Stearns is manufactured in a boiler and fed to an inner pan, it is heated more evenly and avoids localized burns of the product. Viscous foods like sauces, jam, and chutney tend to adhere to the bottom of a pan more easily, making this a critical consideration when reheating them. If this were to happen, the result would be of worse quality as well as substantially more time consuming.

Processing of Fruits and Vegetables

Fermented and distilled products

This set of goods necessitates specialized equipment for fermentation and distillation in addition to the equipment needed to produce juices for fermentation. Either food-grade plastic drums or huge glass vessels with a limited opening, an air lock is used to aid in the fermentation process of wine. The use of an alcohol hydrometer isn't required; however it can help with product standardization by measuring the alcohol concentration. Creating vinegar from wine is conceivable, but the yield is limited and the risk of spoiling is considerable. A professionally constructed vinegar fermented is too expensive for most producers, but a locally created fermented with a conventional design can be made if the expertise is available.

In addition to the high cost of commercial distillation equipment, small-scale producers are more inclined to adopt cheaper, locally created alternatives. These can be used to make acceptable items with the right amount of skill and control over heating. Long, submerged pipes that leave the heating vessel at least 1.5 meters above the liquid level should be installed as a pressure safety precaution. When the still's outlet is blocked, the pressure will not rise to the point of exploding.

Packaging, Filling and Sealing Equipment

A heat sealer can be used to seal any type of plastic film except for uncoated cellulose. Due to the width of the heated bar or wire and the degree of temperature and time control, there are a variety of sealer types available on the market. A bar-type sealer is preferable to a wire-type sealer for dried and liquid goods since they require a wider seal (e.g. 3-5 mm). Additionally, the sealer should feature a thermostat to manage the sealing temperature and an adjustable timer to control the heating time of the seal. To prevent improper sealing, make sure the inside of the package where the seal is to be applied is free of product dust. Pickles and chutneys are typically filled by hand into jars, plastic pots, or bags using scoops or ladles. This is a lengthy process that may necessitate the involvement of a big number of people. Because mechanical fillers for these types of products are prohibitively expensive and typically run at a high throughput, this is the only viable alternative in most small-scale enterprises. A number of liquid fillers are available for use by small-scale businesses, notwithstanding the fact that liquids can also be filled by hand using jugs or ladles. Stainless steel or food-grade plastic tanks can be fitted with gate valves to create gravity fillers. Volumetric fillers and dispensers use a piston to precisely fill each container with a predetermined amount of liquid. It is possible to seal jars and other containers with small devices.

Packaging materials for food come in a broad variety, and it is impossible to cover all of them in a book of this size. Entrepreneurs should contact package manufacturers or their agents for a complete list of available packaging types from the Bibliography. An overview of some of the most often used packaging materials in developing nations may be found below, including a brief discussion of their salient characteristics.

In places where there is a glass-works or an overland supply from a neighboring country, jars and bottles are readily available. To transport glass containers over long distances, the containers must be heavy, large, and fragile; this makes them prohibitively expensive for producers in developing nations. Reusing them is common when they are accessible, although doing so requires tremendous care to ensure that they are thoroughly cleaned. There must be a proper seal between the container and the lid or cork in order for new and re-used containers to be properly sealed. Although the 'Omnia' kind of jar lid is still widely used in many countries, the TOTO type is now the most prevalent. ROPP (roll on pilfer-proof) caps and corks can both be used to seal bottles to prevent them from being stolen. Plastic pots and bottles can be used for a variety of cuisines, and their lower production and distribution costs have led to their rise in popularity. Pots can be sealed with either a foil lid or a snap-on plastic lid by heating them. Polythene and polypropylene are the most popular types of plastic film in underdeveloped countries; however agents who can sell more complex (and expensive) foreign laminate are increasingly available. Many nations now have compact laminated plastic/foil/cardboard cartons for UHT juices, although these are mainly imported under license to large juice manufacturers and are not available to small processors. Because of this, it is not feasible to use UHT technology on a small scale. You'll find a wider variety of packaging options in the form of cardboard and paper. Only in niche export or tourist markets are more traditional forms of packaging like leaves and jute possible to transmit an image of "modern" or "hygienic" products; these are not generally employed.

Legal Aspects of Horticultural Processing

The first step for potential business owners should be to find a professional who can register them as either a sole proprietorship with unlimited liability or as a limited liability corporation with only one shareholder/director. However, if additional partners are needed to contribute finance or specific talents, this may not be appropriate.

In addition, a limited liability company with many directors or an unincorporated group without limited liability can be considered. A co-operative association, a not-for-profit organization, or a registered charity could be the best option if the planned firm has a significant number of interested investors, such as a farmers' association, or if the goals of the venture include social advantages as well. Charity law in several nations, on the other hand, outlaws trading.

Food related laws

Almost everywhere items, including food, are governed by general regulations stating that each product should be acceptable for its intended use. It is illegal to add, process, or sell food that is harmful to health with the goal of selling it for human consumption under the Food and Drug Administration's food rules. Most countries have legislation in place to ensure that consumers aren't duped into buying food that has been tampered with. Typically, they state that it is an offence to sell food that does not meet the purchaser's expectations for its type, substance, or quality. Falsely describing food on the label or in advertising with the goal of misleading the client is also a crime.

Food composition

Composition laws for processed foods are complicated and specialized to certain types of food, such as pies or prepared foods that offer potential for adulteration. There are laws in place to ensure that all items that have the name of a particular food have the same content. Because of the challenges in enforcing this policy, some governments are shifting their stance. Instead, the government is relying on tougher labeling rules to alert consumers of the food's content.

Compositional standards are routinely applied to fruit and vegetable products, such as:

Fruit juices and nectars: When it comes to juice, there should be no additions other than vitamin C, specific acids used to modify the pH, and the maximum quantities of residual Sulfur dioxide that can be found in the juice. Juice should be at least 25 percent and up to 40 percent depending on the fruit, with a maximum of 20 percent sugar or honey in nectars. Nectars must also meet certain acidity requirements.

Fruit juices and nectars if a preservative like sulphur dioxide has been employed, then the juice should only contain pure juice and no additional ingredients like vitamin C, the acid used to regulate the pH, or any other additives like that. Juice should be at least 25 percent and up to 40 percent depending on the fruit, with a maximum of 20 percent sugar or honey in nectars. The acid content of nectars must also meet certain standards.

Soft drinks: Minimum fruit contents for different types of fruit are established for squashes, crushes, and cordials in the law. For drinks that are not diluted, the minimum fruit content ranges from 1.5 percent to 5%, while for those that must be diluted, the required fruit content ranges from 7 percent to 25%. The ratio of water to consume must be four to one. A maximum amount of sugar or artificial sweeteners, as well as certain food acids, are allowed in each.

Jams and similar products: Depending on the fruit used, jams should have a minimum fruit pulp content of 200g/kg of product. Fruit juice content in jellies and jams is also indicated. Sulphur dioxide levels in all food products must be kept to a minimum, and jams must contain at least 60 percent soluble solids by volume. Definitions for jams, jellies, marmalades, conserves, preserves, and extra jam or jelly, as well as reduced sugar jam, jelly, and marmalade, are covered in great depth by laws.

Tomato ketchup: This should contain no more than 6% tomato solids and no seeds. Copper contamination is limited to onions, garlic, and spices for flavoring, and no other fruits or vegetables can be used.

Additives and contaminants: Lists of authorized food colours, emulsifier, stabilizers, preservatives, and other additives are available for use in food products. No chemical can be used that isn't on this list. Lists of goods that can contain certain preservatives, as well as a maximum quantity for each additive in those items, are also in place. Maximum quantities of contaminants, such as toxic metals like arsenic and lead, are allowed in specific foods.

Food Labeling: A vast number of food company convictions are based on 'technical' violations of the law, such as wrongly labeling a product. Label design difficulties and expensive re-design after labels have been printed can be avoided if processors contact the local Bureau of Standards early on. General labeling rules outline what information must be put on a label; however, many countries have laws that go into great depth about any or all of the following:

Specific names for various substances must be specified.

» Customers' capacity to comprehend and access information health-giving or tonic powers, dietary advantages, diabetes or other therapeutic claims that are false and deceptive.

» Guidelines for the use of adjectives like flavour, fresh, vitamin and so on.

Professional guidance should be obtained from graphic designers who are familiar with label design or from a Bureau of Standards.

Hygiene and Sanitation

It is common practice in developing countries to strictly enforce regulations governing the conditions of food producing facilities and the people who work there. If a company isn't in compliance with these rules, food inspectors from health departments or other enforcement agencies can bring legal action against it, and the business can be shut down as a result. Therefore, before submitting a new processing facility for inspection and certification, guidelines on the design and construction of premises and operator hygiene should be consulted.

In order to manufacture products that are both safe and of excellent quality, these standards must be strictly followed during normal production. In a nutshell, health, sanitation, and hygiene are addressed by the legislation:

» Processing that takes place in unclean surroundings.

» In situations where food is at risk of contamination, for example.

» Equipment for food handlers and their obligations to keep the food safe.

» Water supply systems are part of the building design and construction process.

This form of legislation aims to prevent consumers from being duped by dishonest manufacturers, for example by selling them underweight food packets. It is the goal of the rules to ensure that the net weight stated on a label matches the actual weight of food in the package. However, it is acknowledged that not every pack may be filled exactly with the stated weight because of the unpredictability of both machine-filling and hand-filling of containers. As a result of this, policies have been put in place to allow for this variety while preventing fraud. The older and still prevalent in the majority of developing countries, system of weights and measures is known as the Minimum Weight System. At least the weight stated on the label must be present in each package of food. Producers face criminal charges if any of their packages fall below this weight. In order to avoid legal repercussions, producers must routinely fill packs just beyond the specified weight, which results in a small amount of product being given away in every pack.

Automatic filling and packaging, utilized by most European producers, was taken into account by further European legislative measures. Based on a statistical probability that a certain percentage of shipments exceed their claimed weight, this system is known as the Average Weight System (AWS). In poor nations, where most small-scale producers don't use automatic fillers and programmable check-weighers, this technique is difficult to run and unnecessary. A local Export Promotion Board or equivalent agency should be consulted if a producer intends to export to an industrialized country in order to receive the 'e' mark, which indicates that the process adheres to this system.

When selling dried fruits, vegetables, jams, or marmalades, certain countries require that certain weights be used. Many tiny processors receive their daily supply of fruits and vegetables from the public market down the street. For one thing, processors have limited influence over the price charged by traders each day, and because of the huge seasonal price swings that characterize these raw materials, financial planning and cash flow control are made more difficult. Most developing countries' production often falls short of targets because of a lack of adequate planning and resources, as well as the fact that commercial food processing is a relatively new industry with no prior history of collaboration or formal contracts. However, when this has been done, both processors and suppliers benefit, providing that the arrangements are honorable and mutual confidence is in place. As a result, farmers can expect a set price for their crop and a steady stream of customers to buy it when it is ready to be harvested.

There are a variety of reasons why processors should not overlook farmers when negotiating contracts: for example, traders often buy the entire harvest regardless of quality and either sort it themselves or sell it on to wholesalers who perform the sorting. Farmers don't have to worry about marketing their crops or disposing of poor goods because they get paid at the farm. Even though farmers can sell to traders with the knowledge that they would have a 'guaranteed' market, they have little influence over the prices they are offered and are vulnerable to being abused, especially during the height of the growing season when there is an abundance of a specific crop. A number of other services that farmers may not be able to obtain elsewhere are also provided by traders, such as the supply of farming tools and other inputs such as seeds; they are also a source of immediate informal credit, which farmers may need to buy input or for other needs such as funerals and weddings. Despite the fact that the interest rates on these loans may be far higher than those on commercial loans, farmers often have no other option but to take out these loans. Various farmers in many nations are indebted to traders for the rest of their lives, and they can only be freed from the burden by selling their land.

There are several reasons why farmers may not be willing to terminate their current trading agreements with processors, such as genuine fears that they will lose the services they currently receive or the fact that they are indebted to traders and have no other options available. If farmers are unable to get loans, they may be threatened with not buying their crops again, or they may be told they must repay their loans promptly, or they may even be threatened with violence if they sell directly to processors.

Variety, ripeness at harvest, infection-free status, and so on would all be included in a typical specification. Crop prices are determined in advance and can range from the lowest point in mid-season to the highest in pre- and post-season. Alternately, prices can be set according to a sliding scale that takes into account one or more easily quantifiable characteristics, such as minimum size or agreed colour range, in exchange for contracts to supply specific kinds of produce from specific growers or groups of growers who may be working cooperatively.

The price paid for the crop is fixed in advance and might range from the lowest point of the season to the highest point of the season. An alternative is to agree on a pricing scale based on one or more easily measurable criteria, such as minimum size or agreed colour range, with an impartial person present to confirm the agreement if there are later issues. Depending on the agreement, the quantity that can be purchased may also be specified. Even though formal contracts are uncommon in most developing nations, these agreements are typically spelled out in writing and signed by both parties. Additionally, processors should think about how they may help farmers in other ways. Tea and coffee producers, for example, provide training for their workers and an extension service to help them deal with crop problems as they develop throughout the growing season. This may be beyond the means of small scale processors, but more limited support may include purchasing tools, fertilizer or other necessities in bulk and passing the savings on to farmers. Alternatively, farmers can purchase inputs without incurring debt by paying a portion of the crop in advance.

Better control over raw material quality and varieties planted, some control over supply volumes, and an early indication of expected raw material costs benefit the processor. This aids in financial management and production scheduling. Having a guaranteed market for the harvest at a known price, as well as any additional incentives that processors may give, benefits the farmer. In order for this partnership to be successful, both the processor and the farmer must live up to their obligations. They've tried these kinds of agreements on a few occasions, but they've fallen apart due to one person not keeping their end of the bargain. At the end of each growing season, farmers may sell some of their crops to traders for a higher price than they would get from a processor. The processor is unable to handle the anticipated volume of crop and hence cannot meet the intended production capacity, which has a negative impact on sales and cash flow. Alternatively, the processor delays payment to farmers, resulting in the necessity for them to take out another loan and increasing their financial burden. Farmers may also find themselves unable to sell their crops to processors who may refuse to buy them or pay a low price, leaving them with no choice except to look for new markets.

For an additional development, the processor could lease or buy property and build up a separate company to feed the processing unit. When a current farmer diversifies into processing but keeps the farm, this is known as "in reverse" diversification. In either situation, the processor provides all of the farm's labour and supplies. If there's a surplus, it's sold at local markets or by traders.

Agreements with Retailers and other Sellers

Each product in a range has its own marketing strategy, and these decisions are made by processors on how to promote their products and who to sell them to. For example, a jam company may produce two types of jam: one for high-end metropolitan consumers, and the other for bakers who use it in their pastries and confections. Processors must know the market they operate in and how items flow through the market and earn value, regardless of the form of sale they envision.

Each seller needs between 10% and 25% profit to handle, stock, or transport foods, hence less direct routes from producer to consumer result in significant increases in the product's unit cost. Additionally, because wholesalers control a significant portion of the market, producers may need to reduce their profits to supply them; distributors must contribute a higher percentage than any other group to cover high transportation costs in most developing countries; and street vendors and kiosk owners typically make the lowest profits of any group. Direct sales from the processing unit to customers are the simplest and most cost-effective mode of distribution. Bakery products, which have a limited shelf life and are often preferred when fresh out of the oven, are more likely to be sold directly to consumers, but fruit and vegetable products are less likely to be sold this way. Except for the sale of pickles and chutneys from bulk containers into customers' pots and the sales of product packs at the front of the processing unit by a small 'factory shOp' and the sales of fresh, unpasteurized juice at nearby cafes or tea rooms owned by the processor themselves, there are some exceptions.

In these situations, it is necessary to include provisions for client comfort and the availability of sales personnel in the design and layout of the premises. Customers must not be permitted into the processing area in order to maintain control over hygiene, pilferage, and health and safety. If the processing plant is located within a reasonable distance of a significant number of merchants, direct delivery of goods is possible. In order to offset distribution costs, a greater price should be paid by the processor, but retailers can still save money by not purchasing through wholesalers.

In these situations, it is necessary to include provisions for client comfort and the availability of sales personnel in the design and layout of the premises. Customers must not be permitted access to the processing area in order to prevent contamination, theft, and other threats to health and safety.

Direct shipments to retailers are possible if the processing facility is close enough to a large enough population. In order to offset distribution costs, a greater price should be paid by the processor, but retailers can still save money by not purchasing through wholesalers. To cut down on the expense of shipping raw materials, it is common practice to process near a rural area rather than centrally placed industrial units. It is more customary for wholesalers to acquire items from the producing unit or for processors to send products to one or more wholesalers. Proximity of bringing raw materials to a manufacturing site, or of moving finished products to market, should be taken into account during the feasibility analysis of an industrial project. Because of the significant risk of losing both empty packaging en route to the unit and full product en route to wholesalers or retailers, it is important to take extra care while transporting glass jars or bottles across country roads.

Chapter - 8

Storage of
Fruits and Vegetables

In the development of agriculture, the ability to keep harvested plant organs for long periods of time has been crucial. Gathered and stored food was kept in simple baskets as early as 7000 BC. An advanced technology like burned clay pottery would allow early people to store harvests in hidden habitats, generating a simple modified climate. Pre-Neolithic Middle Eastern tribes used underground pits to store grain 9,000–11,000 years ago, while Middle Eastern artists specialised in manufacturing pottery of various forms and sizes for a variety of purposes (4500 BC). Ancient Egyptians and Samarians are believed to have used sealed limestone crypts to keep some of their crops around 2500 BC in order to extend their shelf life. Primitive tribes have used pits to store a wide variety of fruits and vegetables, and they continue to do so. The Romans employed grain silos for long-term storage, and this practise persisted into the nineteenth century. In many cases, the harvesting season for horticultural crops is limited and highly perishable. As a result, preserving these foodstuffs by storing them properly and utilising appropriate techniques will help keep them around for longer. As an added benefit, storing fresh food will assist to reduce oversupply on the market and provide a wider range of fruits, vegetables, and flowers to customers throughout the year, not just during peak season. Fresh food is stored to preserve its freshness and quality, decrease spoiling, increase the usability of the product, and increase the profit to the growers. Physiological processes like as respiration, transpiration, ripening, and other biochemical changes can all be slowed down by the use of storage. Controlling disease transmission and retaining the product's highest quality are further goals of proper storage. The product's basic features and perishability determine how long it can be stored. Products such as raspberries have a limited shelf life, while those such as onions, potatoes, garlic, and pumpkins have a lengthy shelf life. The features of a product influence the storage conditions as well. Leafy vegetables, on the other hand, can withstand temperatures as low as 0o C, although most tropical fruits cannot. Only one crop should be stored in a room at a time unless it is for a brief length of time. The incompatibility of temperature and relative humidity conditions, chilling and ethylene sensitivity, scent contamination, and other issues impacting shelf life and quality might result in product deterioration if the same storage room is used for different items.

Development of storage science

Jacques Berard, a French horticulturist, experimented with manipulating fruit ripening in the 1820s by manipulating the storage environment. While in storage, fruit uses O2 and generates CO_2 and if O_2 is completely depleted, the fruit does not mature. After removing them from the changed gas environment, Berard found that apples, apricot, peaches, pears, and prunes could be preserved for long periods of time in ambient air and room temperature. The evolution of fruit and vegetable storage technologies is greatly influenced by refrigeration. It's no different today than it was in ancient times: refrigeration cuts down on food spoilage and waste by rapidly lowering the temperature of harvested crops. Although ice and snow have been used to preserve food since the Roman era, it wasn't until the 1800s that natural ice from the northern hemisphere was widely employed to chill goods around the world.

Goals of storage

» The drying time and moisture loss of products can be reduced by using this method.

» Minimize bodily harm by avoiding physiological imbalances.

» Maintain the lowest temperature that will not cause freezing or chilling injury by managing the atmospheric composition of the product to slow down its biological activity.

» Microorganisms' growth and spread can be prevented by keeping the product's temperature low and reducing the amount of surface moisture.

» Reduce the product's vulnerability to ethylene damage.

» Prevent market glut and distress sales to extend the market duration.

» Increasing the availability of fruits and vegetables in the off-season.

Classification of fresh horticultural crops according to their relative perishability and potential storage life in air at near optimum temperature and relative humidity

Relative perishability	Potential storage life (weeks)	Commodities
Very high	Less than 2	Apricot, blackberry, blueberry, cherry, fig, raspberry, strawberry, asparagus, bean sprouts, broccoli, cauliflower, green onion, leaf lettuce, mushroom, muskmelon, pea, spinach, sweet corn, tomato (ripe), most cut flowers and foliage, minimally processed fruits and vegetables.
High	2-4	Avocado, banana, grape (without SO_2 treatment), guava, loquat, mandarin, mango, melons (honeydew, crenshaw, persian), nectarine, papaya, peach, plum, artichoke, green beans, brussels sprouts, cabbage, celery, eggplant, head lettuce, okra, pepper, summer squash, tomato (partially ripe).
Moderate	4-8	Apple and pear (some cultivars), grape (SO_2 treated), orange, grapefruit, lime, kiwifruit, persimmon, pomegranate, table beet, carrot, radish, potato (immature).
Low	8-16	Apple and pear (some cultivars), lemon, potato (mature), dry onion, garlic, pumpkin, winter squash, sweet potato, taro, yam, bulbs and other propagules of ornamental plants.
Very low	More than 16	Tree nuts, dried fruits and vegetables

Tips for storage of high quality horticultural produce

» Because many commodities can be damaged by freezing or chilling, only store high-quality food that is free of damage, decay, and is of the right maturity (not overripe or under-mature).

» Storage facilities should be safeguarded from rodents by keeping the adjacent outdoor area clean and clear of rubbish and weeds. • Do not over load storage rooms or stack containers close together and offer appropriate ventilation in the storage room.

» Keep storage rooms clean.

» High humidity settings should be avoided when storing onions and garlic.

» Ethylene-sensitive products should not be stored near those that create ethylene.

» Apples, garlic, onions, turnips, cabbage, and potatoes should not be stored with other foods that absorb odours.

» Prevent the spread of disease by removing any damaged or diseased produce.

Ideal storage temperature

Temperature-sensitive fruits, vegetables, and flowers can't all be stored in the same way. Mould growth and chilling injuries are important considerations to keep in mind, as well as the length of time that must be kept in storage. Reduce the temperature as low as it can be safely handled to extend the shelf life, cut down on waste, and keep quality high during marketing. Fruits, vegetables, and flowers can be stored at temperatures ranging from −1 to 13°C, depending on how quickly they decompose. Some fruits, like apricots, can be stored for up to five weeks at −1 to 4°C, while others like mandarins, nectarines, and ripe or green pineapple can be stored for up to two weeks at 5-9°C. Bananas can be stored between 10 and 13°C for up to one week and green bananas can be stored between 13 and 10°C for up to one week. Asparagus, beans, broccoli, and brussels sprouts can be stored for up to 1-4 weeks at −1- 4° C, while cauliflower can be stored for up to 2-4 weeks at 5-9°C. There are a number of vegetables that can be stored at a temperature of 5-9°C for up to 28 weeks, including non-perishables such as carrots, onions, potatoes, and parsnips. When refrigerated at 10°C, sweet potatoes can last for 16-24 weeks. A correlation between the respiration rate and the storage life of produce has been found to exist, with produce that has a lower respiration rate often lasting longer. The respiration rate of a ripe banana is 200 ml Co_2 $kg^{-1}h^{-1}$ at 15°C, whereas the respiration rate of a non-perishable fruit like an apple is 25 ml Co_2 $kg^{-1}h^{-1}$ at the same temperature.

Principles underlying storability for fruits and vegetables

Fruits and vegetables are living organs:

Respiration is the process through which carbohydrates, organic acids, proteins and lipids are digested. They require oxygen and create carbon dioxide during respiration. Cells get their energy via respiration, which also fuels ripening processes including colour and flavour development. Fruit and vegetable quality is greatly affected by water loss or transpiration. Quality can be affected in numerous ways, including wilting, flaccidness, soft texture and loss of nutritional content, as well as lower saleable weight, due to the loss of water. Losses might range from 3% for lettuce to 10% for onions, depending on the product. Temperature control and humidity control are two ways to prevent water loss from items, as are surface coatings or plastic film in areas where they are permitted. Because active metabolism continues after harvest, storage and shelf life conditions undergo a multitude of positive and unwanted alterations. These changes include development of pigments for example, lycopene synthesis in tomato, anthocyanin synthesis in strawberry, and development of carotenoids (yellow and orange colours) in apricots and peaches. As the fruit matures, it loses chlorophyll (which gives it its green colour) and develops new flavours and aromas. To get rid of chlorophyll in tomatoes is a good thing, but removing it from cucumbers and broccoli is a no-no. Apples benefit from starch to sugar conversion, whilst potatoes do not, and vice versa. Peas and sweet corn, on the other hand, benefit from sugar to starch conversion.

Transpiration losses* for fruits and vegetables stored at various relative humidities

Crop	Storage temperature (°C)	Percent weight loss per day (*% Weight loss per 24 hours, Calculated from ASHRAE data)			
		95% RH	90% RH	85% RH	80% RH
Apples	0	0.011	0.022	0.033	0.044
Brussels sprouts	0	1.610	3.220	4.840	6.420
Cabbage	0	0.058	0.116	0.175	0.233
Carrots	0	0.315	0.630	0.945	1.260
Celery	0	0.460	0.920	1.38	1.840
Table grapes	0	0.036	0.064	0.096	0.128
Leeks	0	0.210	0.420	0.620	0.820
Lettuce	0	1.930	3.860	5.790	7.730
Parsnips	0	0.500	1.000	1.500	2.00
Peaches	0	0.150	0.300	0.450	0.600
Pears	0	0.018	0.036	0.054	0.072
Potatoes uncured	6	0.070	0.141	0.211	0.282
Potatoes cured	6	0.021	0.042	0.063	0.084
Tomatoes	6	0.060	0.119	0.180	0.240

Fruit and vegetable varieties

Marketability and output are important considerations for growers since they immediately impact their bottom line. Resistance to postharvest illnesses and physiological abnormalities should be included in the selection process for new cultivars, as should disease resistance in general. Varying physiologies and biochemistries are to blame for the different storage capacities of different products. The commodity can be developed to have genes that slow down processes such ethylene synthesis, respiration, and softening. There are many commercial tomato varieties that include the rin mutant, for example. Breeders have also favoured choices that ripen more slowly and are therefore more resistant to the abuses that occur during harvest, handling, and transportation, which can lead to bruising and skin damage in the plants.

Maturity at harvest

Any fruit or vegetable, no matter how long-lived, can be significantly affected by harvest ripeness and storage conditions. Commercial harvesting of horticultural crops occurs at the physiological stage known as "flowering." As items age and ripen, they gain desirable consumer attributes like sweetness and flavour. The product's stowing capacity is also decreasing. Fruit that can be stored should be gathered sooner than food that is meant for immediate consumption because of these opposing characteristics. Examples include an apple that is picked at full ripeness, with low starch content, high soluble solids, and is highly aromatic but will have a short storage life, while one destined for long-term storage must be harvested much earlier when the fruit has high starch content and less developed aroma volatiles. Storage life for strawberries that are fully ripe and flavourful is significantly reduced when harvested at the white tip stage of maturation. White tip fruits, on the other hand, tend to be less flavourful than those that are plucked at a later stage.

Pre-harvest management affects storage quality

The mineral content of fruits, vegetables, and flowers during harvest is widely considered to have an impact on their long-term storability. Generally speaking, higher calcium levels are related with longer storage life and greater vulnerability to storage diseases and pathogens. Blossom end rot, a tomato disease, is linked to insufficient calcium concentrations in the fruit. Onion storage rots are often exacerbated by the use of nitrogen, which is commonly used to boost production. Onions with higher nitrogen levels have thicker necks, which are more vulnerable to injury during topping, resulting in faster deterioration. Potassium, magnesium, and boron concentrations, as well as phosphorus deficiency, all contribute to decreased storage stability.

Respiration of the product

In addition to the exchange of gases, water and heat are produced during respiration (energy). A lack of refrigeration or ventilation might cause the commodity to rise in temperature. Inadequate ventilation can allow carbon dioxide to accumulate around the product, and oxygen depletion can lead to fermentation as a result. Storage conditions that reduce oxygen and raise carbon dioxide can be utilised to extend the shelf life of some fruits and vegetables, although this is not always the case. The respiration rate of various types of fruits and vegetables after harvest is generally connected with the rate of deterioration. Product respiration rates can vary widely depending on season, variety, and post-harvest management, but the storage life of most foods is fairly consistent. Foods with high respiration rates like asparagus, mushrooms, parsley, peas, spinach, and sweet corn degrade much more quickly than foods with low respiration rates like apples, beets, celery, garlic, grapes, honeydew melon, and onions. The respiration rates of many fruits and vegetables fluctuate widely, from extremely low to extremely high.

Horticultural commodities classified according to their respiration rates

Class	Range at 5°C (mg CO_2/kghr^{-1})*	Horticultural Products
Very low	Less than 5	Dates, dried fruits and vegetables, nuts
Low	5-10	Apple, beet, celery, cranberry, garlic, grape, honeydew melon, onion, papaya, potato (mature), sweet potato, watermelon.
Moderate	10-20	Apricot, banana, blueberry, cabbage, cantaloupe, carrot (topped), celeriac, cherry, cucumber, fig, gooseberry, lettuce (head), nectarine, olive, peach, pear, pepper, plum, potato (immature), radish (topped), summer squash, tomato.
High	20-40	Blackberry, carrot (with tops), cauliflower, leeks, lettuce (leaf), lima bean, radish (with tops), raspberry, strawberry
Very high	40-60	Artichoke, bean sprouts, broccoli, brussels sprouts, endive, green onions, kale, okra, snap bean, watercress
Extremely high	More than 60	Asparagus, mushroom, parsley, peas, spinach, sweet corn
*Vital heat (Btu/Ton-24 hrs) = mg CO_2/kg hr^{-1} x 220.		

Ethylene production

Ethephon (2-chloroethane phosphonic acid) is a gas that can have effects at concentrations as low as parts per billion (ppb) and as high as parts per million (ppm) (ppm). Ethylene is a naturally occurring component of many fruits' ripening process. Disease and degradation, as well as exposure to cold temperatures and injury, can all boost the body's ethylene production. Internal combustion engines, smoke, and other pollutants also produce ethylene. The respiration of fruits and vegetables can be stimulated by their exposure to ethylene. During maturation and ripening, the respiration rates of non-climacteric fruit gradually decline while in climacteric fruit a sudden increase in respiration is related with the beginning of ethylene production is observed. The ethylene production and sensitivity of apple and pear climacteric fruit is high, whereas the ethylene production and sensitivity of other climacteric produce (e.g. broccoli, cabbage, carrots, and strawberries) can be low, Ethylene production and ethylene sensitivity are minimal in most non-climacteric fruits like cherries, grapes, berries, and peppers, which are not climacteric. When ethylene is exposed to unripe climacteric fruit, it can cause the fruit to ripen earlier than desired, resulting in a mushy, mealy texture. Cucumbers and other green veggies like parsley and broccoli can turn yellow when they're exposed to ethylene gas in the marketplace. Exposed non-climacteric fruits and vegetables also have higher respiration rates, which mean that their carbohydrate reserves are depleted faster and water loss is increased. Ethylene-sensitive and ethylene-producing products are often stored together in the same room, resulting in an ethylene exposure risk. Ethylene can cause the plant to lose its blooms and leaves, as it accelerates the rate of abscission.

Fruits and vegetables classified according to ethylene production rates

Class	Production rate at 20°C (68°F) (μl C_2H_4/kg hr^{-1})	Commodities
Very low	Less than 0.1	Artichoke, asparagus, cauliflower, cherry, citrus fruits, grape, jujube, strawberry, pomegranate, leafy vegetables, root vegetables, potato, most cut flowers
Low	0.1-1.0	Blackberry, blueberry, casaba melon, cranberry, cucumber, eggplant, okra, olive, pepper (sweet and chilli), persimmon, pineapple, pumpkin, raspberry, tamarillo, watermelon

Moderate	1.0-10.0	Banana, fig, guava, honeydew melon, lychee, mango, plantain, tomato
High	10.0-100.0	Apple, apricot, avocado, cantaloupe, feijoa, kiwifruit (ripe), nectarine, papaya, peach, pear, plum
Very high	More than 100.0	Cherimoya, mammee apple, passion fruit, sapota

Internal ethylene concentrations in several climacteric and non-climacteric fruits

Fruits (Climacteric)	Ethylene ($\mu l\ L^{-1}$)	Fruits (Non-Climacteric)	Ethylene ($\mu l\ L^{-1}$)
Apple	25-2500	Lemon	0.11-0.17
Pear	80-100	Lime	0.30-1.96
Peach	0.9-20.7	Orange	0.13-0.32
Avocado	28.9-74.2	Pineapple	0.16-0.40
Banana	0.05-2.1		
Tomato	3.6-29.8		

Ethylene production and sensitivity of several commodities

Commodity	Ethylene production	Ethylene sensitivity
Climacteric fruit		
Apple, kiwifruit, pear, cherimoya	High	High (0.03 - 0.1 ppm)
Avocado, cantaloupe melon, passion fruit	High	Medium (> 0.4 ppm)
Apricot, banana, mango	Medium	High (0.03 - 0.1 ppm)
Nectarine, papaya, peach, plum, tomato	Medium	Medium (> 0.4 ppm)
Vegetables and non-climacteric fruit		
Broccoli, brussels sprouts, cabbage, carrot cauliflower, cucumber, lettuce, persimmon potato, spinach, strawberry	Low	High (0.01- 0.02 ppm)
Asparagus, bean, celery, citrus, eggplant	Low	Medium (0.04 - 0.2 ppm)
Artichoke, berries, cherry, grape, pineapple, pepper	Low	Low (> 0.2 ppm)

Summary of the detrimental effects of ethylene related to quality

Ethylene effect	Symptom or affected organ	Commodity
Physiological disorders	Chilling injury	Persimmon, avocado
	Russet spotting	Lettuce
	Superficial scald	Pear, apple
	Internal browning	Pear, peach
Abscission	Bunch	Cherry tomato
	Stalk	Muskmelon
	Calyx	Persimmon
Bitterness	Isocoumarin	Carrot, lettuce
Toughness	Lignification	Asparagus
Off-flavours	Volatiles	Banana
Sprouting	Tubercle, Bulb	Potato, onion
Colour	Yellowing	Broccoli, parsley, cucumber
	Stem browning	Sweet cherry
Discolouration	Mesocarp	Avocado
Softening	Firmness	Avocado, mango, apple, strawberry, kiwifruit, melon

Storage temperature

A fruit and vegetable grower's most important postharvest tool is temperature management. To slow down the metabolism of fruits and vegetables after harvest, it is essential to lower their temperature as quickly as possible. Reduced respiration rates, water loss, ethylene sensitivity, and decay susceptibility of commodities are all advantages of lower temperatures. Commodities that are stored at lower temperatures tend to last longer. Chilling damage, such as irregular ripening, failure to ripen and pits on the skin surface, can occur when low temperatures are applied to certain commodities. This is why the safest storage temperature for each commodity is not the same. Chilling sensitive products should be stored at a greater temperature than non-chilling products. It is the lowest storage temperature that is associated with the longest storage life in non-chilling sensitive items. It is true that the storage life of chilling-sensitive products rises with decreasing storage temperature, up to a maximum temperature of seven to fifteen degrees Celsius (Celsius). Storage life diminishes as a result of chilling injuries at lower temperatures.

Fruits and vegetables classified according to sensitivity to chilling injury

Sensitivity to chilling	Fruits and vegetables
Non-chilling sensitive commodities	Apples, apricots, artichokes, asparagus, beans (lima), beets, blackberries, blueberries, broccoli, brussels sprouts, cabbage, carrots, cauliflower, celery, sweet corn, cherries, currants, endive, garlic, grapes, lettuce, mushrooms, nectarines, onions, parsley, parsnips, peaches, pears, peas, plums, prunes, radishes, raspberries, spinach, strawberries, turnips
Chilling sensitive commodities	Bean (snap), cranberry, cucumbers, eggplant, muskmelons, peppers, potatoes, pumpkins, squash, sweet potatoes, tomatoes, watermelons, yams

Changes during storage: Several changes such as physical and biochemical occurs during storage.

Physical and biochemical changes during storage

Parameters	Different changes in storage
A. Physical	
Weight loss	Increased. Loss of moisture due to transpiration caused weight loss.
Shriveling	Increased. Even 5 per cent loss of weight result shriveling.
Firmness	Firmness of fruit is due to calcium pectate. Break down of these substances reduced firmness and caused softness in fruits/ vegetables.
Colour	Development in mango, banana, papaya, limes and lemon, etc. Discolouration in litchi, grape and strawberry.
Flavour	In the beginning of storage flavour developed in some fruits such as peach, pineapple, pear and guava and long storage results acidic flavour in many fruits.
B. Biochemical	
Hormones	Level of ethylene increased in climacteric fruits such as mango, pear, apple, banana, etc. Excess of ethylene stimulate softening and pathological disorders in many fruits and vegetables. Abscicic acid (ABA) increased in pear fruit and stimulates ripening of the fruit during storage which cut down the storage life.

Parameters	Different changes in storage
Respiration	Increased level of ethylene during storage increased the rate of respiration which leads to rapid break down of glucose and other organic compounds.
Pigments	Pigments such as chlorophyll and anthocyanin decreased and carotenoids increased in many fruits and vegetables.
Sugar	In the begging of storage increased (slightly break down) then decreased. In table variety of potato increased.
Pectic substances	An increase in pectinic acid and decrease in proto pectin.
Amino acid	Arginin and proline increased e.g. grape but total amino acid content decreased.
Enzyme	An increase in level of amylase, pectin esterase, malicenzyme, catalase, peroxidase and polyphenol oxidase.
Lipids	Palmitic and palmitolitic increased and linoleic decreased in avocado fruits.
Volatile compounds	Ethylene, esters; aldehyde and ketone increased and cause harmful effects.
Phenolic compound	Tanin, flavones and total phenols decreased.
Vitamins	Ascorbic acid (vit. C) decreased and β-carotene increased.

Factors affecting storage life

Pre-harvest factors, maturity at harvest, harvesting and handling techniques, pre-storage treatments, temperature and humidity in storage rooms, and overall hygiene of the produce all affect the storage life of fresh horticulture products. In terms of the aforesaid parameters, temperature and relative humidity are the most crucial ones. Temperature is able to control the physiological and biochemical activity of fresh horticultural crops after harvest. Senescence is accelerated at higher temperatures because of the increased respiration rate. As a result of temperature control, chemical treatments, atmospheric change, and post-harvest disease prevention and control, the storage life of horticultural products can be extended significantly. Low-temperature storage, on the other hand, is the only practical means of preserving horticulture products' quality over the long term. All other approaches are merely meant to supplement the usage of cold storage.

Management of temperature and relative humidity in storage

In order to keep fresh horticulture products safe and fresh for as long as possible after harvest, temperature control is the most effective method available. During the cold chain, it begins with the quick removal of field heat through the use of initial chilling (refrigerated transportation, cold storage at wholesale distribution centers, refrigerated retail display, and cold storage at home). Reducing water loss during storage necessitates temperature and relative humidity control.

Temperature: Maximum shelf life is achieved at temperatures around 0°C for most perishable plants. It takes 2 to 3 times longer to degrade at temperatures above the optimum than it does at temperatures below the optimum. The germination and growth rate of pathogen spores are strongly influenced by temperature, as are other internal and external parameters. Conditions like the following can rapidly deteriorate when temperatures are outside of the ideal range.

Freezing: Water content (75 to 95 percent) and big, heavily vacuolated cells are common characteristics of perishable goods. As a result of their tissues' relatively high freezing point (varying from –3°C to –0.5°C), freezing causes tissue collapse and complete loss of cellular integrity during storage, which is why it is so important to keep them out of the freezer.

Chilling injury: Products from tropical and subtropical regions have a tendency to suffer from chilling injury when stored below a crucial temperature (between 5°C and 13°C depending on commodity and maturation stage), which is known as the chilling threshold temperature or lowest safe temperature for the product. There are many indicators of chilling injury, including as surface and internal discoloration, pitting and water soaking as well as inability to ripen and uneven ripening as well as the formation of unpleasant flavours and an increased sensitivity to pathogens assault.

Heat injury: Perishable goods are likewise severely harmed by high temperatures when they are stored. Direct heat sources like full sunlight can quickly heat tissues to exceed their thermal death point, leading to localised bleaching or necrosis (sunburn or sunscald) or overall collapse if the organs are removed from the plant.

Relative humidity: The vapour pressure difference between the commodity and its surroundings is directly related to the amount of water that is lost. When the relative humidity of the air around a stored commodity is low, the vapour pressure differential will be low as well. Water loss, decay, disease incidence and severity, and uniformity of fruit ripening can all be affected by RH. The RH of the ambient air has less of an effect on deterioration than does moisture condensation on the commodity (sweating). Fruits should be stored at a relative humidity (RH) of 85-95%, but the RH range for the majority of vegetables is 90-98%. 70 to 75 percent relative humidity is ideal for dry onions and pumpkins. RH levels of 95 to 100 percent are ideal for root crops like carrot, parsnip and radish. There are a variety of ways to regulate relative humidity, including adding moisture (water mist or spray, steam) to air via humidifiers; controlling air flow and ventilation in cold storage rooms in relation to produce load; maintaining refrigeration coil temperatures at or near room temperature; installing moisture barriers to insulate storage rooms and transport vehicles; and installing polyethylene liners inside containers (snap beans, peas, sweet corn, and summer squash).

Principles of storage of horticultural fresh produce

Controlling of respiration

The breakdown process of respiration necessitates a storage strategy that minimises this metabolic activity. On the basis of this theory, cold storage, atmospheric modification, and low pressure storage are all viable options. In the centre of the store, the heat generated by respiration is known as respiratory heat. If the heat in the storage chamber isn't eliminated, the respiration rate of the stored produce rises. As a result, a lower respiration rate can be achieved by improved ventilation. The ripening process in some fruits and vegetables can be delayed by reducing the respiration rate.

Controlling of transpiration

After harvest, fresh produce continues to lose water, resulting in wilting or shrivelling. Produce that has lost just 5% of its moisture will shrivel and be unmarketable. Moisture loss from fresh produce is influenced by a combination of factors, including the relative humidity and temperature. Increased storage temperatures will also lead to an increase in water loss. At high temperatures and low humidity, the ripening process of fresh fruits and vegetables is accelerated. As a result, preserving vegetables at low temperatures with high relative humidity can help manage this process.

Controlling of sprouting and rooting

Vegetables that have been sprouted or rooted are not particularly popular with shoppers. As the sprouts dry up, the produce becomes more vulnerable to infection by microorganisms. This type of produce can have its shelf life increased by extending the dormant stage by establishing unfavourable storage conditions. Temperature and humidity levels in the storage area might encourage the growth of microorganisms, which can lead to food degradation. The growth of these microorganisms that cause spoiling should be slowed or controlled by storage measures.

Methods of Fruits and Vegetable Storage

Traditional methods

Prompt storage in a setting that maintains product quality can extend the commercial life of fresh horticulture goods. Temperature, air circulation, relative humidity, and atmospheric composition can all be altered in a variety of environments. Previously, the methods listed below were employed to achieve the desired environmental conditions for long-term storage of fruits and vegetables.

In situ/ on site/ natural or field storage: When harvesting root and tuber crops in the winter season, harvesting is held back until the crop is needed. A wide variety of root vegetables (carrots and turnips), tubers (potatoes and rhizomes) should be left in the ground until they are ready to be sold at the market. Cover root crops with straw, hay, or dried leaves once the ground freezes in the late fall. However, unless the soil temperature drops to -4.0°C or lower, produce will not be harmed by the freezing conditions. Unlike parsnips, carrots begin to deteriorate at a temperature of -4.0°C. New crops cannot be cultivated on the land that has already been used for crops. Citrus and other fruits are often left on the tree in this manner. Cassava's acceptability and starch content suffer, as well as pre-harvest losses, when harvest is delayed. Pests and diseases, as well as freezing and chilling injuries, can harm crops. An underground container or indoor storage location may be necessary if rodents are an issue. Light freezing enhances the flavour of parsnips, horseradish, and turnips. Sugar is formed at temperatures between -2.5°C and 1.1 °C. Early light frosts can be tolerated by other crops, such as beets; cabbage; Chinese cabbage; celery; endive; lettuce; kale; leeks; and onions; however, these crops can be stored for several weeks under heavy mulch if they are properly protected.

Sand and coir storage: In India, potatoes are traditionally stored longer periods of time, which involves covering the commodity underground with sand.

Bulk storage of dried bulb crops: Low humidity is ideal for storing onions, garlic, and dried fruits. Garlic and onions will grow if stored at a temperature between 40°F and 60°F. Sulphur-rich pungent onions can be preserved for longer periods of time than mild or sweet onions, which can only be stored for a few weeks. At 2 cubic feet/minute/cubic feet of produce, ventilation systems should be developed for bulk storage of onions or garlic to bring in air from the bottom of the room.

Storage condition for onion, garlic, dried fruit and vegetables

Commodity	Temperature (°C)	RH (%)	Potential storage duration
Onions	0-5	65-70	6-8 months
	28-30	65-70	1 month
Garlic	0	70	6-7 months
	28-30	70	1 month
Dried fruits and vegetables	<10	55-60	6-12 months

Clamp storage of root and tuber crops: Potatoes in Great Britain and India have traditionally been stored this way. Potatoes that are to be processed should be kept at a moderate temperature in order to minimise the development of darkening sugars that occur when heated. In order to prevent the development of the deadly alkaloid solanine in potatoes meant for human consumption, they must be stored in complete darkness. When storing potatoes for seed, they should be kept in a room with diffused light. Seed potatoes will be protected against decay organisms and insect pests thanks to the accumulation of chlorophyll and solanine. In order to avoid chilling harm, tropical root and tuber crops must be kept at a temperature that is not too low. At the side of the field, the clamp's width is around 1.0–2.5 metres. Potatoes are arranged in an elongated conical stack on the ground and the proportions are marked out. The soil may be covered with straw before the potatoes are planted. Heap height is determined by the angle at which the cluster rests, which is approximately one-third of its breadth. Straw is bent over the ridge at the top of the structure to direct rain away from it. When compressed, straw should have a thickness of 15 to 25 cm. Soil has accumulated to a depth of 15–20 cm after two weeks, however this can vary depending on the weather.

Storage condition for root and tuber crops

Commodity	Temperature (°C)	RH (%)	Potential storage duration
Potatoes (Fresh market)	4-7	95-98	10 months
Seed potatoes	0-2	95-98	10 months

Cassava	5-8	80-90	2-4 weeks
	0-5	85-95	6 months
Sweet potato	12-14	85-90	6 months
Ginger	12-14	65-75	6 months

Pits or trenches: Outside of the field, pits have been dug to collect rainwater. As a rule, pits are usually located near the highest part of the field, especially in areas with a lot of rain. It is lined with straw or other organic material and filled with the crop to be stored, and then topped with a layer of organic material and a layer of soil. Straw is used to make ventilation holes at the top of the crop, as rotting can occur if the holes aren't ventilated.

Zero energy cool chambers (ZECC): For the short-term storage of fruits and vegetables, zero energy cool chambers were developed based on the evaporative cooling principle. The air's original humidity and the evaporating surface's efficiency both play a role in how much cooling is achieved. Brick (400 bricks) and river bed sands (10 sacks) are the primary building materials for cool chambers, each of which can hold around a quintal of food. A single layer of bricks forms the floor of the storage area, which is then covered with a rectangle construction with two walls, with a 7.5 cm gap between the bricks being filled with sand. Outside, the chamber should be 165x115x67.5cm in size. Gunny cloth is used to cover the top of the storage compartment, which is supported by a bamboo frame. Sprinkle water once in the morning and once at night to keep the temperature (with a range of 2 to 5 degrees Celsius) and humidity (90 percent) stable throughout the year. It's best to prevent getting water on your produce in the first place. Crates or matching baskets should be used to store fruits and vegetables before they are placed in the chamber. During the summer, the storage life of many commodities can be extended by up to three times beyond that of the ambient environment.

Storage life of different commodities in zero energy cool chambers

Vegetables	Shelf life (days)		Fruits	Shelf life (days)	
	Ambient condition	ZECC		Ambient condition	ZECC
Leafy veg.	1	2	Ber	5	9
Okra	3	5	Karonda	7	12
Tomato	2	3	Ripe banana	3	5
Brinjal	4	6	Mango	4	6

Bottlegaurd	4	7	Citrus	6	10
Chilli	3	5	Guava	3	5
Watermealon	7	15	Jamun	1	3
Roundgourd	4	6	Lasoda	6	11

Windbreaks: Stakes are driven into the ground in two rows, one metre apart, to create a windbreak. A wooden platform, generally formed from wooden boxes, is constructed between the stakes at a height of 30 cm. Stakes are inserted into the ground and chicken wire is strung across the windbreak at both ends. Onions in the United Kingdom are stored in this manner.

Cellars: A house's foundation typically contains these partially or completely underground rooms. This location offers excellent insulation, so it can keep you cool in hot weather and keep you safe from dangerously low temperatures in winter. During the winter months, apples, cabbages, onions, and potatoes are historically stored in Indian cellars on a small scale.

Ventilation: Using the difference between day and night temperatures in hot climes can help keep stores cooler. When the crop is stored in the room, it should be properly insulated. The store room has a built-in fan that is activated when the temperature outside drops below the temperature within. When the temperature levels off, the fan shuts down. The fan is controlled by a differential thermostat that constantly compares the outside air temperature with the interior storage temperature. Bulk onions and garlic are stored this way.

Natural ventilation: Natural ventilation is the most straightforward of the many available options. Respiratory heat and humidity are removed by utilising the natural airflow around the product. Structures with gaps for ventilation and some degree of protection from the outside environment can be used. Bulk, bagged, boxed, binned or palleted produce is stored. Naturally stored goods are vulnerable to pests and illnesses as well as severe weather conditions, all of which can degrade their quality. Plants like potatoes, onions, sweet potatoes, garlic, and pumpkins that keep well in the ground are the only ones that can benefit from this approach.

Improved storage methods

Various more scientific approaches such as cold storage, controlled atmosphere storage, modified atmosphere storage, solar drive cold storage, hypobaric storage, and jacketed storage are currently being used to achieve the desired environmental conditions for long-term storage.

Cold storage (Refrigeration, Chilling and Freezing):

Low-temperature storage is the most well-known, effective, and extensively utilised method for preserving perishable goods including fruits, vegetables, and flowers while minimising losses and maximising shelf life. Preserving seasonal product and selling it at a profit during the off-season is also made easier by the use of cold storage. As a result of the limited number of cold storage facilities now accessible for a wide range of products such as potatoes, oranges and apples, grapes and pomegranates, flowers, etc., many of these facilities are underutilised.

Number of cold storages and its capacity in states (2018)

S.No	Name of the states	Number	Capacity (In 000MT)
1	Andaman &Nicobar	3	0.810
2	Andhra Pradesh & Telangana	452	1836.36
3	Arunachal Pradesh	2	6.00
4	Assam	37	163.26
5	Bihar	306	1415.60
6	Chandigarh	7	12.46
7	Chhattisgarh	98	484.33
8	Delhi	97	129.86
9	Goa	29	7.70
10	Gujarat	890	3516.00
11	Haryana	352	791.78
12	Himachal Pradesh	65	125.97
13	Jammu & Kashmir	55	182.53
14	Jharkhand	58	236.70
15	Karnataka	209	602.46
16	Kerala	199	81.70
17	Lakshadweep	1	0.015
18	Madhya Pradesh	302	1281.41
19	Maharashtra	603	979.60
20	Manipur	3	7.10
21	Meghalaya	4	8.20
22	Mizoram	3	3.97

23	Nagaland	4	7.35
24	Orissa	177	566.32
25	Pondicherry	3	0.085
26	Punjab	672	2201.39
27	Rajasthan	167	561.30
28	Sikkim	2	2.100
29	Tamil Nadu	173	347.58
30	Tripura	14	45.48
31	Uttar Pradesh	2368	14500.77
32	Uttrakhand	47	162.82
33	West Bengal	511	5940.51
34	NHB/State	3	20.20
	Total	**7916**	**36229.68**

Cold storage capacity in India is estimated at 36.22 million tonnes, with 7916 units in operation. Most of these cold storage facilities can only store potatoes, which only account for around 20% of total agricultural income. Only 10% to 11% of India's fruit and vegetable production is said to be stored in cold storage. To avoid waste, storage capacity must be raised by 40%. Due to the tropical and humid climate, there is increased food waste in India's southern and western areas. In order to boost agriculture's economic impact, India has to build out its cold storage infrastructure in an economical, dependable, and long-term manner.

Factors for Effective Cold Storage of the Produce

Product quality

Produce that is intended for long term preservation should be free from physical damage at the peak of ripeness and free of pathogens.

Temperature: Perishables should be stored at a low temperature in order to slow down respiration and metabolic activity, ageing, softening and colour changes, moisture loss, spoilage owing to diseases and undesirable development, and to prevent spoilage from diseases and unwanted growth (sprouting). It is critical to maintain a constant, uniform temperature and to choose the lowest possible setting for each individual product.

Relative humidity: The quality of vegetables stored in storage facilities is directly impacted by the air's relative humidity. Withering or shrivelling is more likely if the level is too low, while degradation is more likely if it is too high. Relative humidity in cold storage should be maintained between 85 and 95 percent for the majority of fresh fruits and vegetables (with the exception of onion and garlic).

Air circulation and package spacing: This is necessary to remove heat from the lungs, as well as extra CO_2 and ethylene. To maintain a consistent temperature in a cold storage room, the air must be circulated. Additionally, the right spacing of containers on pellets and the introduction of fresh air are critical.

Respiration rates, heat evolution and refrigeration: Respiration is the most critical function of living tissues when a storage facility for fresh products is alive and well. Temperature and commodity type affect how much energy is released in the form of heat throughout this operation. BTU (British thermal units) are used to express this critical heat in terms of the commodity's refrigeration load.

Weight loss in storage: Harvested crops lose a lot of water when they're put into storage. Losses that induce wilting or shrivelling should be avoided at all costs. Moisture loss can be kept under control in good handling conditions with recommended humidity and temperature.

Sanitation and air purification: There is tremendous advantage in good air circulation alone in reducing surface mould. Off flavours and rapid deterioration can both be exacerbated by the build-up of odours and volatiles.

Refrigeration Cycle

Principle of refrigeration:

The evaporator is the first step in the refrigeration cycle. The liquid refrigerant in the evaporator is used to remove heat from the product at this point. The refrigerant is a low-pressure, low-temperature vapour as it exits the evaporator after absorbing heat from the product. The compressor then removes this vapour from the evaporator. Temperature increases as a result of the compression of vapour. Compressors increase pressure by converting low-temperature gas to hotter gas, therefore raising the temperature of the gas. Flowing from the compressor to the condenser, this high-pressure, high-temperature vapour is cooled by the surrounding air or by fan aid. The condenser only cools the vapour to the point when it reverts back to a liquid state. After being absorbed, the heat is subsequently transferred to the ambient air. The expansion valve is now being used to move the liquid refrigerant. The expansion valve lowers the refrigerant pressure, which lowers the temperature of the liquid refrigerant. When the low-pressure, low-temperature liquid refrigerant enters the evaporator through the expansion valve, the cycle is complete. The following elements must be taken into account when determining the refrigeration load:

» Field heat (Temperature of produce at harvest)

» Heat of respiration of the produce

» Conductive heat gain – building floor, wall, roof ceiling etc.

» Convective heat gain – air mixing during opening of door

» Equipment load – fans, lights, forklifts and personnel etc.

» Service and defrost factors of the facility – hot weathers

Components of refrigeration system

Refrigerant is used to transmit heat from a product away from the evaporator, compressor, condenser, and expansion valve in a refrigeration system.

Evaporator: The liquid refrigerant in the evaporator is used to remove heat from the product. Low pressure causes the evaporator's liquid refrigerant to boil. The liquid refrigerant's temperature must be lower than the chilled product's temperature in order to facilitate heat transfer. Refrigerant from the evaporator is drained into the compressor's suction line after it has been transferred. Liquid refrigerant evaporates off of the evaporator coil and escapes as a gas.

Compressor: Low-temperature, low-pressure vapour from the evaporator is drawn into the compressor through the suction line by the compressor. Compression occurs when the vapour is sucked in. Temperature increases as a result of the compression of vapour. As a result, the compressor raises the pressure by converting the vapour from a low- to high-temperature state. The discharge line is where the compressed gas is released into the atmosphere.

Condenser: The condenser's primary function is to transfer heat from the refrigerant to the atmosphere outside the unit. The condenser is usually installed on the reinforced roof of the building, which enables the transfer of heat. Air is drawn through the condenser coils by fans located above the condenser unit. The temperature at which condensation begins is determined by the high-pressure vapour's temperature. As heat has to flow from the condenser to the air, the condensation temperature must be higher than that of the air; usually between -12°C and -1°C. It is then cooled to the point where the high-pressure vapour within the condenser reverts to a liquid refrigerant, retaining part of its heat. The condenser discharges the condensed refrigerant into the refrigerant piping.

Expansion valve: The expansion valve is situated at the end of the refrigerant line before the evaporator in a refrigeration system. The condenser's high-pressure refrigerant makes its way to the expansion valve. The aperture inside the valve reduces the refrigerant's pressure as it goes through the valve. The temperature of the refrigerant drops to a level lower than the surrounding air as the pressure is reduced. The evaporator is subsequently filled with this low-pressure, low-temperature refrigerant.

Low temperature (chilling) injuries

Chilling injury is the primary cause of damage to tropical and subtropical horticultural produce that is stored at low temperatures. Chilling injury occurs when tropical and subtropical fruits and vegetables, such as mango, banana, and tomato, are stored at temperatures higher than their freezing point but lower than 5°C to 15°C, depending on the product. Chilling injury happens when an object's surface temperature falls below a specific threshold. There is a big difference between a chilling injury and a freezing injury (which results when ice crystals form in plant tissues at temperatures below their freezing point). Chilling injury is product and even cultivar specific in terms of vulnerability as well as symptoms. In addition, the same commodity grown in different regions may respond differently to the same temperature conditions.

Symptoms of chilling injury

Skin pitting: Cells beneath the skin have collapsed, leading to this typical cold injury sensation. It is common for the pits to be discoloured and for water to be lost from damaged regions, highlighting the extent of pitting.

Browning or blackening of flesh tissues: Affected fruits, such as avocado, usually show signs of chilling-induced browning first on the vascular (transport) strands. During cooling, phenolic molecules released from the vacuole may be activated by the enzyme polyphenoloxidase (PPO), which may cause browning. However, this mechanism has not been verified in all situations.

Rotting: The release of metabolites (e.g. amino acids, carbohydrates, and mineral salts) and mineral salts from cells is caused by a chilling damage. Pathogenic microorganisms, such as fungi, benefit from the breakdown of cell membranes, which allows metabolites and ions to leak out. During harvest and postharvest operations, these pathogens may be present as latent infections or contaminate produce.

Other symptoms: Slow de-greening of citrus fruits the formation of an off-flavour or aroma (low O_2 levels), and immature plucked fruits may fail to ripen or ripen unevenly or slowly after chilling injury are all examples of different systems (e.g. tomato). Pineapple has a black heart, a woolly peach and a brown plum. Physiological diseases such as these are also common. Chilling injury symptoms typically appear when the product is at a cool temperature. When the produce is transported to a higher temperature, chilling injury may manifest and degeneration can occur quickly, frequently within a few hours.

Lowest safe storage temperature and symptoms of chilling injury in produce

Produce	Lowest safe storage temperature (°C)	Symptoms
Apple	2-3	Soft scald, brown core
Avocado	5-12	Pitting, browning of pulp and vascular strands
Banana	12-13	Brown streaking on skin
Ber	7.5-8.5	Surface pitting, appearance of brown streaks on the peel
Guavas	4-5	Pulp injury, decay
Lemon	10-12	Pitting of flavedo, membrane staining, red blotches
Lime	7-8	Pitting
Mango	12-13	Dull skin, brown areas
Melon	7-10	Pitting, surface rots, browning, water loss, decay
Papaya	7-15	Pitting, water-soaked areas
Pineapple	6-15	Brown or black flesh
Asparagus	4-5	Gray tips and loss of sheen
Beans	5-10	Opaque discolouration and rusty spots
Casaba	7-10	Pitting, surface decay, failure to ripen
Cucumber	7-8	Dark-coloured, water-soaked areas, water loss, decay

Eggplant	7-8	Surface scald, pitting and brown spotting
Okra	7-8	Discoloration, water-soaked areas, pitting, decay
Onion	15-18	Water soaked scales
Pepper	7-8	Sheet pitting, alternaria rot on pods and calyxes, darkening of seed
Potatoes	3-4	Sweetening
Squashes	12-13	Pitting, water loss and decay
Sweet Potatoes	12-13	Decay, pitting, internal discoloration; hard core when cooked
Tomato ripe	10-12	Water soaking and softening decay
Tomato mature green	12-13	Poor colour when ripe, Alternaria rot
Pumpkins	15-18	Pitting, water loss and decay, especially alternaria rot
Watermelons	4-5	Pitting, objectionable flavour

Management of Chilling Injury

Maintaining critical temperature:

In order to prevent chilling harm, it is best to identify the product's critical temperature and avoid exposing it to temperatures below that threshold.

Modified atmosphere storage:

As a result, gases are removed or introduced to alter the atmosphere around the item, creating an atmosphere distinct from that of air (78.08 % N_2, 20.95 % O_2 and 0.03 % CO_2) typically; this is accomplished by reducing O_2 levels and/or increasing CO_2 levels.

Maintaining high relative humidity:

Chilling injury signs, such as pitting, can be minimised by maintaining a high relative humidity during and after storage at a low temperature (e.g. film-wrapped cucumbers).

Mechanism of chilling injury

An integrated genotypic property of the organ, the critical temperature below which chilling harm occurs is represented in phenotypic characteristics. Primary and secondary cellular events of chilling injury can be categorised. The start of cell death and necrosis makes primary events reversible for a short period of time, but they become irreversible after that point.

Primary events in chilling injury

Due to changes in the physical state of membrane lipids, low temperatures alter the characteristics of cell membranes (membrane phase change). By disrupting the synthesis of reactive oxygen species like hydrogen peroxide, enzyme activity and structural proteins like tubulin are both harmed.

Secondary events in chilling injury

The physical changes in membrane lipids alter the properties of their parent membranes.

» Membrane-bound enzymes' activities are interrupted, causing ion and metabolite movement across the damaged membranes.

» Disruption of sub cellular compartmentation is a common symptom of membrane disturbance and can be observed as an increase in ion leakage from tissues.

» Enzyme activity changes can lead to an unbalanced metabolism, which can lead to cell death.

» Buildup of hazardous substances (e.g. acetaldehyde).

» Chiller-sensitive tissues cause structural proteins of the cell cytoskeleton (such as tubulin) to dissociate.

Effect of cold storage on subsequent behaviour of horticultural produce

The process of ageing and degradation is slowed when food is stored in a refrigerator. Because storing depletes the product's life expectancy, stored goods cannot be kept as fresh as when they were first collected. The fruit or vegetable may need to be properly ripened after storage in some circumstances. To prevent quality loss due to perspiration, raise the temperature of refrigerated storage produce gradually. Due to varying temperature needs, certain produce cannot be stored with others. If the optimal low temperatures do not differ greatly, it may be necessary to keep various products together on occasion. In mixed storage, it's best to avoid the cross-contamination of odours, ethylene, and strongly scented produce. Produce can be preserved without causing any harm because of its compatibility.

Compatibility groups of fruits and vegetables

Group	Temperature	Crops	Status of commodities
Group 1	0-2°C and 90-95% RH	Apple, apricot, asian pear, grapes, litchis, plum, prunes, pomegranate, mushroom, turnip, peach	Produce ethylene
Group 2	0-2°C and 95-100% RH	Asparagus, leafy greens, broccoli, peas, spinach, cabbage, carrot, cauliflower, cherries	Sensitive to ethylene
Group 3	0-2°C and 65-70% RH	Garlic, onions dry	Moisture will damage these crops
Group 4	4-6°C and 90-95% RH	Cantaloupes, guava, mandarin, tangerines	-
Group 5	8-10°C and 85-90% RH	Beans, potatoes (with cipc treatment), cucumber, brinjal, okra, pepper. potato, beans, okra, peppers, olive	Sensitive to chilling and ethylene
Group 6	13-15°C and 85-90% RH	Mangoes, banana, tomato ripe, grapefruit, coconuts, guava, limes, papaya, pineapple, tomato, pumpkin, ginger	Sensitive to chilling injury
Group 7	18-21°C and 85-90% RH	Pears, tomato, watermelon, yams and sweet potato	-

When transporting and storing a variety of fruits and vegetables, compatibility is determined by the following factors:

Temperature: Fruits and vegetables are generally divided into two groups:

» Commodities such as apples, grapes, lettuce and pears that are not sensitive to chilling (such as -2.0 to -0.5°C) and can be maintained up to 2°C above their freezing point

» Produce that might be damaged by heat should be stored between 5°C and 15°C to preserve its freshness and flavour, depending on the variety, stage of ripeness and length of time it will be in storage.

Relative humidity (RH): In the absence of a few exceptions, fresh food should be stored around 90-95 percent relative humidity in order to reduce water loss. You can mix the dried goods with other produce kept at 90-95 percent relative humidity if they are packaged in moisture-proof containers. Water condensation on product should be avoided as a means of cutting down on decay rates.

Ethylene: There are many fruits that release ethylene gas, which can be harmful to ethylene-sensitive products such apples, peaches, bananas, and tomatoes (such as avocado, broccoli, cabbage, carrot, kiwifruit, lettuce, persimmon and watermelon). The yellowing of broccoli and cucumber, the russeting of lettuce, the softening of fruits, the bitterness of carrots, the tissue maceration in watermelons, and the abscission of eggplant calyx are all symptoms of ethylene-induced diseases. Commodities susceptible to ethylene must be protected from ethylene exposure at every stage of the postharvest handling process. Elements such as ethylene can be removed from the storage facility's circulating air to less than one part per million (ppm) in order to enable the distribution and sale of ethylene-sensitive goods.

Odour volatiles: Don't mix garlic, onion, pepper and potato with anything that can absorb these scents (such garlic, leek or onion) (such as apple, avocado, citrus fruits, grape and pear).

Sulfur dioxide: In order to prevent *Botrytis cinerea*-caused deterioration, some table grapes are transported with SO2 generator pads attached to them. Grapes should be stored in a separate container from other fruits and vegetables, as SO2 may wreak havoc on them.

Smart fresh (1- methylcyclopropene): The ethylene action inhibitor Smart Fresh (1-MCP) can be mixed with untreated commodities because 1-MCP does not transfer from the treated to the untreated products.

Organic produce: In order to eliminate pesticide residues from direct contact, organically cultivated produce should be handled and stored separately from conventionally farmed product. Maintaining a log of the cleaning products and dates used by those who handle organic produce is a legal requirement.

Cold chain

When food is grown in the field, it is pre-cooled, transported, stored, distributed, and stored in the home at a low temperature (refrigeration) until it is ready to be eaten. When a farmer harvests his or her crops they must be graded for export or local trade, packed in containers, and then delivered in refrigerated vehicles to a cold storage facility or wholesale market in refrigerated containers. Temperature control is essential for wholesale markets. In order to keep its freshness from harvest till consumption, produce that has been precooled after harvest should never be exposed to temperatures that are too high or too low.

Optimum cold storage conditions and approximate storage life of fruit and vegetables

Fruits/ Vegetables	Temperature (°C)	RH (%)	Approximate storage life (weeks)
Fruits			
Apple	0-2	85-90	20-30
Apricot	-0.5 to 0	90-95	1-3
Artichoke	0-1	95-100	2-3
Chilling tolerant varieties avocado	4.4-5.0	85-90	4-5
Chilling sensitive varieties avocado	12.5-13.0	85-90	2-3
Cavendish green banana	12-13	85-90	3-4
Cavnedish ripe banana	12-13	85-90	1-5
Ney Poovan green banana	12-13	85-90	2-3
Ney poovan ripe banana	8-10	85-90	1-2
Barbados cherry	0-1	85-90	7-8
Ber	5-6	85-90	4-5
Black berry	-0.5 to 0	90-95	2-3 days
Blue berries	-0.5 to 0	90-95	2-3
Cherries Sweet	-1 to 0.5	90-95	2-3
Cherries Sour	0-1	90-95	1-2
Coorg mandarin (main crop)	8-9	85-90	8-9
Coorg mandarin (rainy season)	8-9	85-90	6-7
Sathgudi orange (Moosambi)	8-9	85-90	15-16
Custard apple	15-16	85-90	1.5-2
Coconut	0-1.5	80-85	4-8

Date	6-7	85-90	2-3
Fig	1-2	85-90	6-7
Grapes	-0.5 to 0	90-95	2-8
Grapefruit	10-15	85-90	6-8
Guava	10-12	85-90	2-5
Jackfruit	11-12	85-90	6-7
Jerusalem artichoke	-0.5 to 0	90-95	17-21
Kiwi fruit	-0.5 to 0	90-95	13-21
Lemon	10-13	85-90	4-26
Lime	9-10	85-90	6-8
Litchi	2-3	85-90	8-10
Mandarin	4-7	90-95	2-4
Mango	12-13	90-95	2-3
Alphonso, (Mango mature green)	12-13	85-90	4-5
Banganapalli, (Mango mature green)	12-13	85-90	5-6
Melon (Others)	7-10	90-95	2-3
Nectarine	-0.5 to 0	90-95	2-4
Olives, fresh	5-10	85-90	4-6
Orange	0-9	85-90	8-12
Papaya green	9-10	85-90	3-4
Papaya turning	9-10	85-90	2-3
Passion fruit	6-7	85-90	3-4
Peach	-0.5 to 0	90-95	2-4
Pear	-1.5 to 0.5	90-95	8-30
Persimmon	-1 to -0.5	90-92	13-17
Pineapple (green)	9-10	85-90	4-6
Pineapple (25% yellow)	6-7	85-90	1-2
Plum	-0.5 to 0	90-95	2-5
Pomegranate	7-8	85-90	10-12
Quince	-0.5 to 0	90-95	8-13
Raspberries	-0.5 to 0	90-95	2 days
Sapota mature	20-22	85-90	1-2

Strawberry	0-1	85-90	1-2
Water melon	10-15	90-95	2-3
Vegetables			
Asparagus	0-2	95-98	3-4
Snap beans	8-10	85-90	3-4
Winged beans	10-12	85-90	8-10
Bean (dry)	4-10	40-50	26-43
Beetroot	0-1	90-95	8-10
Brinjal	10-11	90-95	1-2
Broad beans	0-2	90-98	1-2
Broccoli	0-1	95-100	3-4
Cabbage(wet season)	0-2	90-95	4-6
Cabbage(dry season)	0-2	90-95	12-13
Cactus leaves	2-4	90-95	2-3
Cantaloupe (half slip)	2-5	95-98	2-3
Cantaloupe (full slip)	0-2	95-98	1-2
Capsicum(green)	7-8	85-90	3-5
Carrot topped	0-2	90-95	20-24
Carrot (bunched)	0-1	95-100	1-2
Cassava	0.5-1.0	85-86	4-8
Cauliflower	0-2	90-95	7-8
Celery	0-2	90-95	7-8
Chicory	0-1	95-100	2-3
Chinese cabbage	0-1	95-100	8-13
Coriander leaves	0-2	90-95	4-5
Chow chow	12-13	90-95	2-3
Cucumber	10-11	90-95	1-2
Garlic(bulbs) dry	0-1	65-70	28-36
Ginger	8-10	75-80	16-20
Gourd, bottle	8-9	85-90	4-6
Gourd, snake	18-20	85-90	1-2
Green onions	0-1	95-100	3-4
Horse radish	-1 to 0	98-100	43-51

Kohlrabi	0-1	98-100	8-13
Leek	0-1	95-100	8-13
Lettuce, leaf	0-1	95-98	0.5-1
Lima bean	3-5	95-98	0.5-1
Mushroom	0-1	95-98	1.0-1.5
Muskmelon, Honey dew	7-8	85-90	4-5
Okra	10-11	90-95	1.0-1.5
Onion, Red	0-1	65-70	20-24
Onion, white	0-1	65-70	16-20
Pea, green	0-1	90-95	2-3
Parsley	0-1	95-100	4-8
Parsnip	0-1	95-100	17-26
Pepper (bell)	7-13	90-95	2-3
Potato (early)	7-16	90-95	1-2
Potato (late)	4.5-13	90-95	21-42
Pumpkin	12-15	70-75	24-36
Radish, topped	0-1	90-95	3-5
Squash	12-15	70-75	8-24
Sweet Potato	10-12	80-90	13-20
Sweet corn	0-1.5	95-98	0.5-1
Spinach	0-1	90-95	10-14
Summer squash	5-10	95-98	1-2
Taro	7-10	85-90	17-21
Mature green tomato	12-13	85-90	4-5
Red ripe tomato	5-6	85-90	1-2
Turnip	0-1	90-95	8-16
Watermelon	12-15	80-90	1-2
Yam	16-20	60-70	8-30

Cooling concept

Heat, as a kind of energy, continually tries to maintain balance. Field heat can be rapidly removed by an appropriate cooling technology that removes at least 78% of the heat. The 7/8 cooling time refers to the amount of time needed to remove 7/8 of the field heat. For shipments to distant markets, it is highly suggested that 7/8 of the field heat be removed during cooling, and this may be done in a relatively short period of time. Removing the final 8% of the field heat is possible with minimal impact on the product throughout the subsequent refrigerated storage and handling stages. There must be a rapid evacuation of field heat so that the temperature can be brought down. The field heat from the product is transferred to the cooling medium when cooling is used. Temperature, timing, and contact all affect cooling efficiency. The product must be in the pre-cooler long enough to remove heat and achieve maximal cooling. During the cooling process, the cooling medium (air, water, crushed ice) must be kept at a constant temperature. As a result, the chiller must maintain constant, close contact with the surface of each vegetable.

Different methods of cooling

Cooling methods vary depending on the type of product and the size of the company. The vast majority of growers and storage operators use room cooling as their primary technique of cooling their crops and storing their products. Fruit and vegetable growing regions use a variety of cooling methods, including forced air, icing, hydro-cooling, and vacuum chilling. Temperature, time, and contact with the commodity all affect cooling rate, but the cooling method also plays a role. The heat-removal capacity of different cooling media is determined by the type of product being cooled. Using a cold storage room's typical cooling curve (Fig. 93), we can see the notion of "half-cooling" or "seven-eighths-cooling" times in practise. It is the time it takes to bring the product's temperature down from when it was first put in the cold room to the room's specified temperature. To put it another way, half cooling is the time it takes for the product temperature to drop from 20°C to 10°C or 15°C while the room temperature is set at 0°C or 10°C, respectively. This time period will be repeated to reduce product temperature by half again. In other words, seven-eighths cooling is three times as long as half-cooling.

Cold storage management

One of the most challenging components of running a small farm is controlling the temperature. Refrigerators, in contrast to huge apple storages, are generally scarce amongst many farmers; hence, the ideal circumstance of more than one cold storage unit is difficult to find. As a result, improper storage of fruits and vegetables is quite common. Chilling injury can occur if products are stored at temperatures lower than those suggested for safe storage, whereas a product's shelf life will be reduced if temperatures are raised. 0-2°C is the lowest temperature where ethylene sensitive and high ethylene producers are found. Low temperatures inhibit ethylene generation and activity, making it less likely that harm may occur during storage. While it is ideal to keep ethylene concentrations below 1 ppm, in some cases, such as when apples are present, this may be impossible, and the recommended seven-day safe period may be shortened. No of the length of time, fruit and vegetable storage should be done separately.

When designing a cooler, selecting the right cooling unit is critical. Ice will build up on the condensers if the temperature difference between the air and the cooling unit is too large. Many fruits and vegetables can be dehydrated by this drying of the air. In order to avoid water evaporation, a humid environment is required. Dry goods coolers are more frequent and less expensive; although they can only maintain a relative humidity (RH) of 90 to 95 percent at 0°C. Cool Bot™ is a new technology that may be useful for smaller farmers who need precise temperature control for their fruits and vegetables, but mechanical refrigeration is still the most common option. Rooms without mechanical refrigeration units can be made cold with this technology, which modifies the air conditioner's lower temperature limit. In order to keep ice from forming on the evaporator fins of the air conditioner unit, the controller functions. A tiny heater and two temperature sensors are employed. To deceive the air conditioner into continuing to cool below its typical set point, the heater is attached to a temperature sensor on the air conditioner. In order to cool down the air conditioner sensor and shut off its compressor, the Cool Bot uses a first sensor to gauge room temperature. Once that temperature has been reached, the heater in the Cool Room is turned off. The evaporator fins are coupled to the controller's secondary sensor, which can detect the presence of ice. To prevent the compressor from running while ice builds up, a controller shuts down the heater until the ice melts. High relative humidity (RH) reduces water loss in stored perishables because the controller keeps it from cooling to below its preset frost point, allowing any condensation on the fins to be recycled back into the cold room air. The room must be well-insulated and free of air gaps in order for the air conditioning equipment to function properly.

Minimizing water loss during storage

95% of the water in fresh produce, flowers, and other edibles can be attributed to their firmness and crispness, and 95% of the water in these products can be attributed to their metabolic processes, cannot restore lost water after harvesting. As a result, it is critical to minimise post-harvest water loss in order to preserve product quality and sellable weight. Evaporation causes water to evaporate from fresh vegetables. Products with a vast surface area (such as leaves) lose water at a significantly higher rate than products that are compact and spherical (e.g. potato). A waxy cuticle or abscission layer or even pubescent hair can also assist limit water loss from the skin. Also, if these natural barriers are cut or abrasions are made, water loss will rise. As RH drops, fresh food loses more water more quickly. As a result, it is advised that RH be kept between 85% and 95% in the majority of cases. Warmer air carries a lot more water vapour than colder air, so this is another key point to keep in mind. When the temperature rises to 10°C, the air at 0°C with 100% RH will have only around 50% RH. Temperature swings in commodities and the atmosphere can be minimised by reducing water loss during transport and handling.

» In the early hours of the morning, handle fresh products for storage and immediately cool commodities and keep them in good condition.

» Natural tissues that restrict water loss at the side of wounds can be developed by curing/drying specific tuber and root vegetables as well as bulbs, citrus fruits, and tubers.

» As an additional layer of protection, apply waxes and other surface coatings.

» With the use of humidifiers, ensure that the RH is maintained at the highest possible level without causing the degradation of commodities or containers.

» The evaporator coils should operate at a temperature of 0°C or less, depending on the cooling system design.

Importance of cold storages

The importance of cold storage for fruits and vegetables has to achieve following objectives.

Seasonal production: Fresh fruits and vegetables are in high demand throughout the year, but their supply is limited to the season. Jammu and Kashmir and Himachal Pradesh's apples, for example, are sold throughout the country. As a result, cold storage facilities are critical to the long-term storage and distribution of fresh produce.

Spoilage: One of the most significant issues with fruits and vegetables is the large losses caused by contamination. It is estimated that 20-33% of the entire production is lost due to spoiling during the various stages of marketing. In order to reduce current spoiling losses, cold storages must be implemented as quickly as possible.

Losses in transit: Fruits and vegetables are highly perishable and can't be stored for long under normal storage conditions hence they are prone to spoilage during transport. Using refrigerated trucks and air-cooled waggons to transport goods will help reduce losses.

Better distribution and more affordable prices: Consumers pay the producer 40-60% less than the producer's cost. Terminal markets in cities like Mumbai, Chennai, Delhi, and Calcutta, among others, are notorious for fruit and vegetable fraud. A significant portion of the fresh fruit sold in these markets is imported. With improved distribution, the manufacturer never learns what the price of his goods was at the nearest cold stores, and so receives a fair price.

Stabilizing market prices: In addition to the duty of distributing on a demand and time basis and also stabilising market pricing. Aside from providing farmers with the option to produce cash crops at reasonable pricing, cold storage facilities stabilise market prices for perishable fruits and vegetables by supplying them at low or no cost to consumers.

Maintain quality: Cold storage aids in the preservation of product quality for an extended period of time. Perishable commodities can be stored for a lengthy period of time thanks to this method.

Reduced wastage: Using a cold storage system can significantly cut down on the amount of perishable commodities wasted. The distribution of fruits and vegetables is hampered by the absence of a well-organized supply chain. Cold storage and processing extend the shelf life of produce so that it can be consumed outside of peak season.

Controlled Atmosphere Storage (CA Storage)

One of the most cutting-edge methods of food preservation is the CA storage of fruits and vegetables. In Canada, W. R. Philips came up with the idea. In order to store Co_2 in a controlled atmosphere, it is necessary to keep the O_2 concentration below 8% and the CO_2 concentration above 1% (approximately 78 percent N_2, 21 percent O_2, and 0.03 percent CO_2). In order to preserve the freshness and safety of fresh fruits, ornamentals, vegetables, and their products throughout postharvest handling, the use of atmosphere modification should be considered.

Physiological basis of CA storage: 20.9 per cent O_2 and 78.1 per cent N_2 make up the majority of air, while the remaining 0.003 per cent CO_2 and trace amounts of Ne, He, CH_4, and water vapour make up the rest. The ripening process is slowed and respiration rates are reduced in CA storage because oxygen is reduced and CO_2 is elevated.

Biological basis of CA effects: Horticultural crops can experience a stress reaction when exposed to low O_2 and/or elevated CO_2 atmospheres that are within the range of tolerance for each commodity, but when exposed beyond this range respiration and ethylene production rates can be enhanced. As a result, the body is more susceptible to physical illness and death. Low oxygen, physical or chemical damage, and exposure to temperatures, RH, C_2H_4 concentrations outside the optimum range for the commodity all contribute to elevated CO_2 induced stressors.

Temperature, duration of exposure to stress-inducing concentrations of O_2 or CO_2, and fruit maturity and ripeness stage (gas diffusion properties) influence the transition from aerobic to anaerobic respiration. In the absence of oxygen (fermentative metabolism), fruits and vegetables can partially recover and resume their regular respiratory metabolism in the presence of oxygen (respiration). Insecticidal atmospheres (1% O_2 and/or 40 to 80 percent CO_2) and fungistatic atmospheres (>10 percent Co_2) stress plant tissues, yet plant tissues are resilient. Reduced O_2 or elevated CO_2 levels have a greater effect on post-climacteric fruits, reducing their tolerance and reducing their potential to recover. There are many factors that influence recovery, including the length and intensity of stressors as well as metabolism-driven cell repairs in the body's cells.

Elevated CO_2 atmospheres reduce the activity of ACC synthase, a key regulatory site of ethylene biosynthesis, while low CO_2 and/or low O_2 levels increase the activity of ACC oxidase, a key regulatory site of ethylene biosynthesis. Elevated CO_2 levels slow down ethylene's reaction. Biosynthesis of carotenoids (yellow and orange colours), anthocyanins (red and blue colours), and phenolic chemicals are all slowed by optimal atmospheric components (brown colour). Vegetables become tougher due to the slowing down of cell wall disintegrating enzymes and enzymes involved in lignification in controlled atmospheres. Acidity, starch to sugar conversion, sugar interconversions and biosynthesis of flavour volatiles are all affected by low O_2 and/or high CO_2 atmospheres. Ascorbic acid and other vitamins are better preserved when produce is maintained in an ideal climate. This results in improved nutritional quality. There is an increase in pyruvate dehydrogenase activity in the presence of severe stress CA circumstances, while pyruvate decarboxylase is stimulated or activated. Because of this, when products are subjected to stress CA circumstances that are above what they can tolerate, compounds like lactate, alcohols, and ethyl acetates build up, which can be harmful to the commodities. Cultivar, maturity and ripeness stage, storage temperature and length (and in some cases ethylene concentrations) all influence the specific reactions to CA. Argon or helium can improve the diffusivity of O_2, CO_2, and C_2H_4, but they have no effect on plant tissues directly and are more expensive than N_2 as a CA component. N_2 is an inert component of CA storage.

It is possible that ethylene-induced de-greening in non-climacteric commodities and fruit ripening in climacteric crops may be hastened by super-atmospheric levels of O_2 up to roughly 80 percent (such as scald on apples and russet spotting on lettuce). Oxygen poisoning occurs when the concentration of oxygen in the air exceeds 80%. Super-atmospheric O_2 levels in California are likely to be used only in conditions where they can mitigate the detrimental effects of fungistatic, elevated CO_2 atmospheres on commodities that are sensitive to CO_2 induced harm.

Essential features of CA storage

» -1 to 3 degrees Celsius can be maintained with mechanical refrigeration.

» There is no gas leakage in the CA storage chamber.

» A cylinder of nitrogen gas is used to inject the gas into the storage once the room has been filled and sealed.

» CO_2 can be removed from the atmosphere using dry hydrated lime and various CO_2 scrubbers such as ethanolamine, aluminium calcium silicate, activated carbon, magnesium oxide, and magnesium oxide.

» Crops have their own unique requirements for atmospheric composition. However, 2-5 percent oxygen and 3-10 percent carbon dioxide are the most prevalent gas mixtures. CO_2 and O_2 levels in the storage room's air are being monitored on a daily basis.

Beneficial effects of CA (Optimum composition for the commodity):

If O_2 levels fall below 8 percent and/or Co_2 levels rise above 1 percentage point, the ethylene sensitivity of fruits and vegetables can be reduced. Assist in preventing or alleviating physiological problems like chilling damage to various commodities, lettuce russeting, or storage problems like apple scald.

Senescence (and related biochemical and physiological processes), including ripening, softening and compositional shifts, can be slowed by slowing down respiration rates.

After-harvest pathogens (bacteria and fungus) and decay rates can be affected by CA in either a direct or indirect way. Strawberries, cherries, and other perishables are much less susceptible to botrytis rot when exposed to CO_2 concentrations of 10 to 15 percent.

Low oxygen (1%) and/or excessive carbon dioxide (40-60%) can be effective in controlling insects in several fresh and dried fruits, flowers, vegetables and nuts.

Limitations of CA storage

»   Causes specific physiological problems, such as black heart in potatoes, brown staining of lettuce, brown heart in apples and pears, and chilling injury of tropical commodities.

»   Fruits such as banana, mango, pear, and tomato might ripen unevenly if they are exposed to low O_2 levels or high CO_2 levels for more than two to four weeks.

»   When there is a lack of oxygen and high CO_2 concentrations, anaerobic respiration results in the development of off flavours and off odours (as a result of fermentative metabolism).

»   When gas is not available on a timely basis, expensive and technological know-how is needed.

»   Fruit that has been physiologically damaged by low O2 or high CO_2 concentrations is more prone to rot.

Commercial application of CA storage

In recent years, there have been a number of improvements to CA storage that have resulted in better quality control. nitrogen can be created by separating compressed air through molecular sieve beds or membrane systems, and other methods include storing it with low O_2 (1.0–1.5%), storing it with low ethylene (under 1 L L-1), storing it with rapid CA (as quickly as possible to achieve the optimal levels of O_2 and CO_2), and storing it with programmed (or sequential) CA. Additionally, new technologies for establishing, monitoring, and maintaining CA using edible coatings or polymeric films with the appropriate gas permeability to create a desired atmosphere around and within the commodity may expand the use of atmospheric modification during transport or distribution. Fresh-cut food is frequently packaged in MAP (modified atmosphere packaging). The *Botrytis cinerea* degradation and fungistatic CO_2 levels damage flower petals and/or the stem and leaves connected with them, limiting postharvest life of cut flowers. Anti-ethylene compounds, on the other hand, are more cost-effective than CA in reducing ethylene action in flowers. There is a greater commercial application of CA storage on apples and pears around the world than on other crops and fruits such as cabbages, sweet onions and kiwifruit. As a result of long-distance travel, atmospheric modification is employed to alter the flavour and texture of apples, asparagus, bananas, kiwifruits, mangos, peaches, nectarines, plums, and strawberries. Increasing the number of applications for fresh horticulture commodities and their products will require future technology improvements to deliver CA at a fair cost (benefit/cost ratio).

Classification of horticultural crops according to their CA storage potential at optimum temperatures and RH.

Storage duration (months)	Commodities
>12	Almond, Brazil nut, cashew, filbert, macadamia, pecan, pistachio, walnut, dried fruits and vegetables
6 to 12	Some cultivars of apples and European pears
3 to 6	Cabbage, Chinese cabbage, kiwifruit, persimmon, pomegranate, some cultivars of Asian pears
1 to 3	Avocado, banana, cherry, grapes, mango, olive, onion (sweet cultivars), some cultivars of nectarine, peach, plum and mature-green tomato
<1	Asparagus, broccoli, cane berries, fig, lettuce, muskmelons, papaya, pineapple, strawberry, sweet corn, fresh-cut fruits and vegetables, some cut flowers

Recommended conditions for storage of some fruits and vegetables under CA storage

Commodity	Temperature Range (°C)	CO_2 Range (%)	O_2 Range (%)
Apple	0-5	1-2	2-3
Banana	12-15	2-5	2-5
Strawberry	0-5	15-20	10-11
Kiwi fruit	0-5	4-5	2-3
Nuts and dry fruit	0-25	0-100	0-1
Tomato (Mature green)	12-20	0	3-5
Tomato (Partially ripe)	8-12	0	3-5
Lettuce	0-5	0	2-5

Modified Atmosphere Storage (MAS)

"Modified Atmosphere" storage (MA) refers to the storage of products in an atmosphere with reduced oxygen levels and increased carbon dioxide concentrations, often to a level of 21 percent in ambient air. Product respiration or active gas injection can be used to develop MA, which can be either passive or active (MA packaging, MAP). As a result, factors such as product kind, temperature, permeability of the plastic film, and product to bag ratio all play a role in the atmosphere in the bags. With MA storage, gas concentrations in the atmosphere around a product are less well-controlled. The sole difference between the MA and CA is that the CA is more precise. The development of polymeric films with a wide range of gas permeability has sparked interest in the creation and maintenance of altered environments within flexible film packaging.

Biochemical and physiological basis of MA

Every 10 degrees Celsius increase in temperature causes a doubling of respiration and metabolism. As a result, lowering the temperature, increasing the CO_2 level, or some combination of these can all be used to reduce respiration. CO_2 and O_2 levels each have a distinct impact on the rate of respiration. As a result, the overall impact could be either additive or synergistic. The rate of breathing slows down when the oxygen concentration falls below a critical 10%. Anaerobic respiration, on the other hand, can occur when oxygen content falls below 2%, resulting in the accumulation of ethanol and acetaldehyde. Lower pH in plant tissues is also a factor that has a beneficial effect in the growth of plant tissue. Ethylene biosynthesis and action are similarly affected by the loss of water.

Modified storage atmosphere condition and storage temperature

Commodity	Temperature range °C	Modified Atmosphere	
		Per cent O_2	Per cent CO_2
Apple	0-5	1-2	0-3
Apricot	0-5	2-3	2-3
Avocado	5-13	2-5	3-10
Banana	12-16	2-5	2-5
Blackberry	0-5	5-10	15-20
Blueberry	0-5	2-5	12-20

Cherimoya and atemoya	8-15	3-5	5-10
Cherry (sweet)	0-5	3-10	10-15
Chilli	5-12	3-5	0-5
Cranberry	2-5	1-2	0-5
Custard apple	12-20	3-5	5-10
Durian	12-20	3-5	5-15
Fig	0-5	5-10	15-20
Grape	0-5	2-5 (5-10)	1-3 (10-15)
Grapefruit	10-15	3-10	5-10
Kiwifruit	0-5	1-2	3-5
Lemon and lime	10-15	5-10	0-10
Litchi	5-12	3-5	3-5
Mango	10-15	3-7	5-8
Nectarine	0-5	1-2	3-5
Nuts and dried fruits	32-50	0-1	0-100
Olive	5-10	2-3	0-1
Orange	5-10	5-10	0-5
Papaya	5-15	2-5	5-8
Peach (Clingstone)	0-5	1-2	3-5
Peach, freestone	0-5	1-2 (4-6)	3-5 (15-17)
Pear, Asian	0-5	2-4	0-3
Pear (European)	0-5	1-3	0-3
Persimmon	0-5	3-5	5-8
Pineapple	8-13	2-5	5-10
Plum	0-5	1-2	0-5
Pomegranate	5-10	3-5	5-10
Rambutan	8-15	3-5	7-12
Raspberry	0-5	5-10	15-20
Strawberry	0-5	5-10	15-20
Asparagus	0-5	Air	5-10
Bean green	5-10	2-3	4-7
Broccoli	0-5	1-2	5-10
Brussels sprouts	0-5	1-2	5-7

Cabbage	0-5	3-5	5-7
Cantaloupes	2-7	3-5	10-20
Cauliflower	0-5	2-5	2-5
Cucumber	8-12	3-5	0
Leek	0-5	1-2	3-5
Lettuce	0-5	2-5	0
Mushrooms	0-5	3-21	5-15
Okra	8-12	3-5	0
Onion (green)	0-5	1-2	10-20
Onion (bulb)	0-5	1-2	0-10
Peas	0-5	2-3	2-3
Pepper (bell)	8-12	3-5	0-5
Potato	4-12	None	None
Radish (topped)	0-5	1-2	2-3
Spinach	0-5	Air	10-20
Sweet corn	0-5	2-4	10-20
Tomato (green)	12-20	3-5	3-5
Tomato (partially ripe)	8-12	3-5	0
Tomato (ripe)	10-15	3-5	3-5

Environmental factors affecting MA storage

Temperature and relative humidity: Commodity temperatures are influenced by the temperature of the surrounding air. Additionally, the film's permeability, which rises in direct proportion to temperature, is affected by temperature shifts. The permeability of CO_2 is more responsive than that of O_2. Most film packages have low permeability regardless of the relative humidity in the air. Moisture and vapour are well-protected by most common films because they maintain a high internal humidity even in dry environmental conditions.

Light: The photosynthesis of green vegetables, which uses a considerable quantity of CO_2 and reduces O_2, would interfere with the respiration process, which is necessary to maintain a specific MA level in the package. If light isn't excluded, potato quality can suffer. Such products necessitate the use of opaque packaging.

Sanitation factors: Plant pathogens may thrive in the high humidity found in MA packages, which may encourage their proliferation. As a result, proper cleanliness and conditions that encourage the growth and reproduction of microorganisms must be taken into consideration. Thus, the use of fungicides in the packaging of vegetables is critical.

Differences between CA and MA storage

CA Storage	MA Storage
High degree of control over gas concentration	Low degree of control over gas concentration
Longer storage life	Less storage life
More expensive technology	Less expensive technology
Atmosphere is modified by adding gas	It is created by either actively (addition or removal of gas) or passively (produce generated)
Specific temperature should maintain	May or may not be maintained

Maximum storage life (days) in normal atmosphere storage (NA), controlled atmosphere (CA) and low-pressure storage (LP)

Commodity	Maximum storage time (days)		
	Normal atmosphere storage	Controlled atmosphere	Low-pressure storage
Apple	200	300	300
Asparagus	14-21	Slight benefit-off odours	28-42
Avocado (Lula)	30	42-60	102
Banana	14-21	42-56	150
Carnation (flower)	21-42	No benefit	140
Cherry (sweet)	14-21	28-35	56-70
Cucumber	9-14	14+ (slight benefit)	49
Green pepper	14-21	No benefit	50
Lime (Persian)	14-28	Juice loss, peel thickens	90
Mango (Haden)	14-21	No benefit	42
Mushroom	5	6	21

Papaya (Solo)	12	12+ slight benefit	28
Pear (Bartlett)	60	100	200
Protea (flower)	<7	No benefit	30
Rose (flower)	7-14	No benefit	42
Spinach	10-14	Slight benefit	50
Strawberry	7	7+ off flavour	21
Tomato (mature-green)	7-21	42	84

Limitation of Modified Atmosphere (MA) and Controlled Atmosphere (CA) Storage

The anaerobic compensation point occurs when a reduction in oxygen concentrations near fruits and vegetables lowers respiration rates until a substantial increase in respiration rates occurs. When acetaldehyde and ethanol build up in harmful concentrations during this respiration, damage occurs. If the carbon dioxide content in the atmosphere is too high, fermentation can cause similar damage. Across the globe, CA storage is used primarily for apples. CA's use is restricted to a small number of products, despite the fact that it has a major impact on the quality of many other products. In order to alter the atmosphere, structures must be airtight and chilled, with perfect temperature control. In order to get the most bangs for your buck, you'll need a lot of space in these vaults. Long-lived plants that can be preserved for months rather than days or weeks are needed to get the best return on investment (ROI). Furthermore, the return on investment from extending the storage life is frequently insufficient. If a product has a week's worth of storage, for example, adding a few extra days isn't usually a big deal. CA storage tents with lower cost oxygen and carbon dioxide monitors are now available, which may be more suited for shorter-lived products than traditional CA storage systems. The tents and controls have been successfully utilised for local market strawberries, for example, and are stored in cold storage rooms.

Hypobaric storage, in which commodities are stored at low atmospheric pressures, has the promise for non-chemical storage, however practical, cost-affordable solutions are not yet available. Commercially accessible is Dynamic CA (DCA), which reduces the oxygen surrounding goods to levels approaching to the anaerobic compensation point, when fermentation occurs. Fluorescence signals or ethanol concentrations in the fruit are used to monitor the fruit's metabolism, and oxygen concentrations are regulated to prevent harm.

Products which are incompatible in long-term storage

Products			Effects
Apple or Pears	with	Celery Cabbage Carrots Potato Onion	Ethylene from apples and pears damages or causes off flavours in vegetables
			Potatoes cause earthy flavour in fruit.
			Potatoes are injured by cold temperatures
			High humidity causes root growth in onions.
			Ethylene causes bitterness in carrots
Celery	with	Onions or carrots	Odour transfer occurs between products
Meat, egg and dairy product	with	Apple and citrus	Fruit flavours are taken up by the meat, eggs, and dairy products.
Leafy greens and flowers	with	Apple, pear, peaches, tomatoes and cantaloup	Ethylene produced by the fruit crops damages leafy greens and flowers.
Cucumbers peppers and green squash	with	Tomatoes Apples Pears	Ethylene from tomatoes, apples, and pears causes loss of green colour. This is aggravated by storage temperatures of 7-10°C which are too warm for apples and pears.

Fruits and vegetables classified according to their tolerance to low O_2 concentrations

Minimum O_2 concentration tolerated (%)	Commodities
0.5	Tree nuts, dried fruit and vegetables
1	Some cultivars of apples and pears, broccoli, mushrooms, garlic, onion, minimally processed fruits and vegetables .

2	Most cultivars of apples and pears, kiwifruit, apricot, cherry, nectarine, peach, plum, strawberry, papaya, pineapple, olive, cantaloupe, sweet corn, green bean, celery, lettuce, cabbage, cauliflower and brussels sprouts
3	Avocado, persimmon, tomato, peppers, cucumber, artichoke
5	Citrus fruits, green pea, asparagus, potato, sweet potato

Fruits and vegetables classified according to their tolerance to elevated CO_2 concentrations

Minimum CO_2 concentration tolerated (%)	Commodities
2	Apple (Golden Delicious), Asian pear, European pear, apricot, grape, olive, tomato, sweet pepper, lettuce, endive, chinese cabbage, celery, artichoke and sweet potato
5	Apple (most cultivars), peach, nectarine, plum, orange, avocado, banana, mango, papaya, kiwifruit, cranberry, pea, chilli, eggplant, cauliflower, cabbage, brussels sprouts, radish, carrot
10	Grapefruit, lemon, lime, persimmon, pineapple, cucumber, summer squash, snap bean, okra, asparagus, broccoli, parsley, leek, green onion, dry onion, garlic, potato
15	Strawberry, raspberry, blackberry, blueberry, cherry, fig, cantaloupe, sweet corn, mushroom, spinach, kale, Swiss chard

Solar driven cold stores

The refrigeration cycle in tropical climates makes use of sun energy. For the storage of 10 tonnes of bananas (capacity 50 m2), single stage ammonia/water absorption refrigerators with 13 kw peak cooling power were created in Sudan and were designed to maintain a minimum 5°C temperature. In comparison to typical cold storage facilities powered by electricity, this system is more expensive.

Jacketed storages

The refrigeration system intercepts and removes heat from the floor, walls, and ceiling before it reaches the storage area in these double-walled storages. Cooling surfaces can be found on the walls, ceiling, and floor. The relative humidity is kept at or near 100%. The cost of these Canadian-made jacketed storages is 10% higher than that of ordinary storages.

Low pressure storage / hypobaric storage

Low pressure of 0.2 – 0.5 atmospheric pressure and temperature of 15–24°C in an airtight chamber are ideal conditions for storing fruits. Sucking air and generating a vacuum lowers the pressure. The effective O_2 concentration decreased from 21% to 2% when the presser was adjusted from 1 atm to 0.1 atm. If there is too much pressure on an apple, it doesn't cause it to ripen. Volatiles such as CO_2, acetaldehyde, acetic acid, ester, and others that have been released from storage are also eliminated or decreased. It appears that hypobaric storage is limited to high-value produce such as cut flowers because of the high expense of building hypobaric storage facilities. It is also impossible to manage the gases while they are being stored.

Comparative storage life (in days) of produce stored in refrigeration and under hypobaric conditions.

Fruits	Cold storage	Hypobaric storage	Vegetables	Cold storage	Hypobaric storage
Pine apple (ripe)	9-12	40	Green pepper	16-18	50
Grapefruit (ripe)	30-40	90-120	Cucumber	10-14	41
Strawberry (ripe)	5-7	21-28	Beans	10-13	30
Sweet cherry	14	60-90	Onion (green)	2-3	15
Banana (unripe)	10-14	90-150	Lettuce	14	40-50
Avocado (unripe)	23-30	90-100	Tomato (mature green)	14-21	60-100
Apple (unripe)	60-90	300	Tomato (breaker stage)	10-12	28-42
Pear (unripe)	45-60	300			

General recommendations for storage: The University of California (Thompson *et. al.*, 1999) recommends three combinations of temperature and relative humidity

Temperature (°C)	RH (%)	Crops
0 – 2	90 – 98	Leafy vegetables, crucifers, temperate fruits and berries
7 – 10	85 - 95	Citrus, subtropical fruits and fruit vegetables
13 - 18	85 – 95	Tropical fruits, melons, pumpkins and root vegetables

Note: ethylene level should kept below 1 ppm during storage

Five different storage conditions for fruit and vegetables storage are as 0°C temperature and 90-100 per cent RH, 7-10°C temperature and 90-100 per cent RH, 13°C temperature and 85-90 percent RH, 20°C temperature and 80-85 per cent RH and ambient conditions.

Influences of pre-harvest factors on storage quality

It is important to consider pre-harvest conditions when determining the quality of the crop at harvest, as well as its storage and nutritional potential. Fresh horticultural products' postharvest quality is highly dependent on the quality they reached when they were harvested in the first place. Many pre-harvest factors are known to affect storage quality, including genotype and cultivar selection, stage of maturity when harvested, temperature, light intensity, rainfall amounts, soil texture and fertility, fertiliser type and application rates, disease and insect pressure, application of growth regulator and pesticide. As the growing season progresses, it becomes increasingly difficult to determine which pre-collection factors have the greatest impact on post-collection fruit quality. While managing pre-harvest parameters, the ultimate goal is to harvest the crop at its maximum quality and maintain that quality for the duration of storage. Fruit, vegetable, and flower harvests all involve mechanical stress and the removal of plant parts from their parent plants, which results in a reduction in available resources such as water, nutrients, hormones, and energy. Stresses might have an effect on the plant after it has been harvested. Every plant system is at risk from stress because it can disrupt, impede, or accelerate the body's regular metabolic processes in a negative way. However, post-harvest considerations are challenging because many storage regimens benefit from stress conditions such as temperature and environment alteration to improve fresh crop storage capacity. Therefore, it can be argued that pre-harvest conditions have a significant impact on both the state of the crop at harvest and the crop's storage and nutritional potential.

Potential risks of microbial contamination and recommended preventive measures

Production step	Risks	Prevention
Production field	Animal fecal contamination	Avoid animal access, wild, production or even pets.
Fertilizing	Pathogens in organic manure	Use inorganic fertilizers. Proper composting
Irrigation	Pathogens in water	Underground drip irrigation Check microorganisms in water
Harvest	Fecal contamination	Personal hygiene. Portable bathrooms. Risk awareness
Field transportation	Pathogens in containers and tools	Use plastic bins. Cleaning and disinfecting tools and containers
Pack house	Fecal contamination	Personal hygiene. Sanitary facilities. Avoid animal entrance. Eliminate places may harbor rodents.
	Contaminated water	Alternative methods for precooling. Use potable water. Filtration and chlorination of recirculated water.
Storage and transportation	Development of microorganisms on produce	Adequate temperature and relative humidity. Cleaning and disinfection of facilities. Avoid repackaging. Personal hygiene. Do not store or transport with other fresh products. Use new packing materials
Sale	Product contamination	Personal hygiene. Avoid animal access. Sell whole units. Cleaning and disinfection of facilities. Discard garbage daily.

Chapter - 9

Preparing Product from Market

Prepared fruits and vegetables are needed for sale after the harvest. Retail, wholesale, or supermarket chains, as well as farms can all engage in this. Preparation for the fresh market is a four-step process, regardless of the destination:

» Removal of unmarketable material.

» Sorting by maturity and/or size.

» Grading.

» Packaging.

This means that any working arrangement that cuts down on handling will save money and improve quality. As a result, market preparation is best done in the field. Tender or perishable products or small quantities for neighboring marketplaces are the only ones that make this feasible. For large operations, remote or demanding markets, or items requiring particular operations like washing, brushing, waxing, controlled ripening, refrigeration, storage, or other specific treatment or packaging, products must be delivered to a packinghouse or packing shed.

There is no need to choose between these two systems. In many cases, some of the field preparation is done in the packing shed after the harvest has taken place. Primary selection of fruits and vegetables is always done in the field because handling unmarketable units is a waste of time and money. Defective products can be removed from the market in this manner.

Field preparation for lettuce involves three employees cutting, preparing, and packing the lettuce. Field-prepared boxes are transported to pack houses for palletizing, pre-cooling and occasionally cold storage before delivery to distant markets. These boxes are then shipped. Large volumes can be handled in a brief amount of time with the help of mobile packing sheds. A mobile grading and packing line is fed by harvest crews. It is replaced with an empty truck when the consignment has been shipped to its final destination.

Harvested goods are transported to a packing facility where they are packaged and ready for distribution. For primary selection in the field, harvest crews frequently employ an inspection line.

THE PACK HOUSE

Special operations can be carried out in a packinghouse. Products may be prepared 24 hours a day, regardless of the weather, which is another benefit. Farmers' groups, cooperatives, and even community organizations can take use of these possibilities because of their ability to process big volumes.

For example, if you're processing crops for your own use, you'll need a larger packing shed than if you're providing service to others. Packing sheds can be as simple as a complex as a fully-automated warehouse. In other circumstances, packing sheds are supplemented with commercial sales offices and storage spaces.

The term "pack house" refers to a facility that protects both the product and the workers from the elements. A central handling operation ensures that all products are prepared at the same time. To a certain sense, this is like a factory assembly line, where raw materials from the field are processed and packaged into finished goods.

Packing house design

To be effective, a packinghouse must be near the producing region and easily accessible via major roads or highways. It's also important to have a single entry to make delivery and supply easier to manage. There should be room for future expansion or the construction of additional facilities. Enough outside space is also necessary to prevent traffic jams caused by vehicles coming and exiting. The loading and unloading areas of buildings should be built to provide adequate shade for the most of the day. In the summer, they need proper ventilation, and in the winter, they need protection.

It is common for packinghouses to be constructed with low-cost materials.

However, it is essential to establish a pleasant working and living environment for everyone involved. As a result, the quality of a product that is exposed to unfavorable conditions might rapidly decline. Unpleasant working circumstances might also lead to unnecessary rough treatment of employees.

The loading and unloading of goods should be made easier by the presence of ramps in a packinghouse. Forklifts should be able to enter and exit through large doors and open areas. One working day's worth of product should be stored in the reception area. As a result, the packinghouse will remain open in the event that product flow from the field is interrupted (rain, machine breakdown, etc).

The appliances, refrigerators, and most importantly, lighting we use every day could not function without electricity. Lighting (both intensity and quality) is crucial in finding flaws on inspection tables because pack houses typically work long hours or even constantly during the harvesting season. To avoid eye strain and glare, lights should be placed below the level of the user's eyes. It's best to keep light intensity between 2000 and 500 lx for light-colored items, and between 4000 and 5000 lx for darker ones. There should be illumination in both the office and the rest of the building. This is to avoid momentary blindness when the eyes are lifted due to contrasts generated by shaded areas. Equipment, conveyor belts, and clothing must all have dull colours and non-glossy surfaces. Defects aren't hidden by light reflection this way. It also helps to alleviate eye strain.

Washing trucks, containers, and equipment as well as dumping need a steady supply of water. Hydro cooling may also be required in specific circumstances. It is just as important to have a good waste water disposal system as it is to have a decent source of drinking water. If at all possible, administrative offices should be high and situated in pristine, peaceful surroundings. This is for the sake of transparency. Quality control facilities and laboratories should be available in packing houses. After figuring out the features of the building plan, it is essential to produce a diagram for the movement of product throughout the packinghouse and the tasks to be conducted for the entire operation. Product movement should always be in a single direction, with no crossovers, and handling should be kept to a minimum. Concurrent operations, such as working on multiple sizes or maturation stages at the same time, may be possible.

Packing house operations

Reception: There should be a minimal amount of time between harvest and delivery of a packed product. The goods should be shielded from the light as much as possible at reception, where delays are common. Samples for quality analysis are sometimes taken before the product is weighed or counted before entering the facility. Especially when delivering a service to other manufacturers, it is important to keep records.

Dumping onto packinghouse feeding lines is the first step in preparing for the fresh market. There are two ways to dump: dry and wet. To minimize harm and regulate product flow, drop decelerators are essential in both scenarios. Free-floating fruits can be moved by submerging them in water to prevent bruising. Wet goods, on the other hand, aren't suitable for all. Other items, such as salts (such as sodium sulphate), are diluted in water to increase flotation because their specific density is lower than water's.

Removal of rejects: After dumping, one of the first steps is to remove any unusable materials. That's because it's expensive to handle plant debris that won't be used for anything but compost. This is done before sizing and grading is done. In the process of preparing a market for sale, one of the four essential procedures is primary selection. Overgrown, undersized, decaying or malformed units are removed in this phase of the process. Mesh screens, pre-sizing belts, or chains are commonly used to remove very small produce. Hand removal is the standard method for removing damaged, rotting, abnormally shaped, or yellowed plant parts. In many crops, soil and loose portions are removed by brushing and brushing is used to remove the dry leaves clinging to the bulbs of garlic and onions. Differential floatation can be used to separate rejects in crops that can be dipped in water. Soil, latex, insects, pesticides, and other contaminants can be removed with detergents and brushes. In order to preserve fresh fruits, sponges or hot air should be used to dry them.

Bruised, rotten, and other plant pieces that are removed from fruits during the harvesting process, such as culls, can be fed to livestock. In spite of their high water content, they provide an excellent source of energy and are incredibly flavorful. As a result, they have a lower nutritional value than other forms of food. Their low protein and dry matter content is why this is the case. To avoid stomach issues, they must be included in the diet in the correct proportions. Another drawback is that many of them are perishable and hence cannot be preserved for long periods of time. As a result, they can't be introduced to an animal's diet over time. They can be utilized as hygienic fillers or organic soil amendments if they are not used for animal feeding. Before or after sorting by colour, the pack house performs sizing, which is a fundamental function. Grading should always be preceded by both of these procedures. As a result, units with flaws are more easily identified on a homogeneous product, whether in terms of size or colour. A weight-based system and a dimension-based approach are two of the most common options (diameter, length or both). Grapefruits, oranges, onions, and other spherical or nearly spherical goods are among the easiest to sort by size. There are a variety of methods available, including-

Expanded separation distances between mesh filters and diverging belts and rollers Rings of specified diameter can also be used to perform sizing manually. Weight-sensitive trays are used in various crops for sorting purposes. All units of the same mass are automatically transferred to a separate belt using these devices.

In terms of importance, this is perhaps the most critical of the four basic functions. It is the process of categorizing products into different quality levels or categories. There are static and dynamic systems. Crops with a high tenderness or monetary value tend to have static systems. Sorters eliminate any units that don't fulfill the grade or quality category's requirements once the product is placed on an inspection table. In most cases, the dynamic system is the more common. Sorters eliminate defective units as the product goes along a conveyor belt in front of them. The main flow has the best quality. Second and third-grade quality units are frequently removed from their belts and transferred to other belts. In terms of volume sorted per unit of time, it is far more efficient. However, it's important to ensure that all employees are well-trained. This is due to the fact that each unit is only visible to the worker for a brief period of time. Both removing high-quality products from the main flow and failing to remove low-quality products are regular blunders. A second or even third quality rating is provided by rejects based primarily on aesthetic considerations. These can be sold in lower-end shops or used as a raw material for further processing and manufacturing processes.

In contrast, small-scale processing has to meet or exceed the quality standards of major enterprises. In other cases, this is not possible due to the fact that industrial operations often use specific types and processing methods. Surpluses for the fresh market, as well as substandard products, do not supply uniform sources of raw materials. Because of the poor industrial yield and the low level of technology used in the production process, the final product may be of varied quality. As a reminder, the quality of a processed product is directly related to the quality of the raw materials used and how they are processed.

Special operations

Commodity-specific actions are required for these tasks. They are distinct from basic operations in that they are performed on every harvest regardless of the size and sophistication of the packinghouse.

Colour sorting

These are frequent in fruits and fruit vegetables, and can be done by means of a computer programme. Within a certain range of maturity, fruits are often collected for sale. Reduced colour sorting can be achieved by harvesting at the same stage of maturity. Low-volume enterprises, on the other hand, can benefit from this strategy.

Waxing

Waxed fruits like apples, cucumbers and citrus fruits like peaches and nectarines are used for a variety of purposes, including preventing the fruit from drying out, extending the fruit's shelf life after harvest, and sealing minor wounds that occur during handling. Some fungicides are transported by waxes, and others are employed just to enhance gloss and aesthetics. A wide variety of waxes in various forms and compositions can be found.

Sprays, foams, immersion and dripping are all possible methods of application. Even distribution is crucial. Using soft brushes, rollers, or other ways, the product is applied evenly and thoroughly to the fruit's surface. Fruit gas exchange can be obstructed and tissue hypoxia can result if the application is too heavy. Characteristics like as internal darkening and development of off-flavors and smells are among them. The approval of waxes for human consumption is critical.

De-greening

Prior to harvest, the weather has a significant role in the development of greening. Fruits like citrus, for instance, often reach commercial maturity with epidermal flecks of green (flavedo). As with fruits having colour, customers can tell they aren't ripe enough and haven't developed their full flavour. Chlorophyll breakdown is used in de-greening to reveal natural pigments that had been hidden by the green colour. Citrus fruits are exposed to an environment containing ethylene (5-10 ppm) for 24 to 72 hours (depending on the degree of greening) in specially constructed chambers with high relative humidity (90-95%). De-greening is a process that is unique to the production location. Oranges should be kept at 25-26°C, grapefruit and lemon at 22-24°C and mandarins at 20-23°C, according to Artes Calero.

Controlled ripening

Quality and shelf life are directly related to a crop's maturity at harvest. In order to avoid bruising and losses during shipping, fruits must be harvested a little early when they are exported to foreign markets. The ripening process needs to be accelerated and uniform before to distribution and retail sales. So that products can reach customers when they are ready, this is one of the primary reasons. Ethylene is utilized in larger amounts than for de-greening. Bananas are a good illustration of how this sort of activity works. During controlled ripening, the temperature and relative humidity of the room are controlled, and ethylene is eliminated when the process is complete. To get the pulp to the right temperature, you have to heat it up first. After that, the appropriate concentration of ethylene is injected. In these circumstances, the product is kept for a period of time and then ventilated to remove any collected gases. To ensure safe transit and storage, the temperature is lowered after the treatment is complete. Temperature affects ethylene concentration and exposure time, which speeds up the process.

Pest and Disease control

Postharvest treatments are used to keep pests and illnesses under control. Citrus, apples, bananas, stone fruits, and other fruits have all been treated with fungicides from a variety of chemical families. Fungistatic activity is a common property. The spores may be inhibited or reduced, but the disease may not be completely halted. Most commonly, we see the application of chlorine and sulphur dioxide.

Chlorine is the most often used sanitizer, according to the CDC. There's no way to stop a disease that has already taken hold. Fumigation with sulphur dioxide at a concentration of 0.5 percent for 20 minutes is commonly used to control post-harvest infections in table grapes. During storage, 0.25 percent concentrations of fumigation are performed every 7 to 10 days. Inside packaging, sodium metabisulfite-coated pads can be used to prevent the growth of bacteria. These produce sulphur dioxide when they come into touch with fruit's moisture.

Exterminating all stages of insect life with gas fumigation is the gold standard for pest control. Most countries have outlawed the use of methyl bromide as a fumigant after decades of widespread use. High and low temperature treatments, controlled atmospheres, alternative fumigants, or irradiation have replaced this kind of fumigation.

Chemical therapies can also be used to avoid various post-harvest physiological issues. Dips or sprays with calcium chloride (4-6%) can be used to treat apples that have a bitter pit. As an alternative, chemicals can be used to prevent scalding or other problems in the fruit during storage. While 2.4-D can help keep citrus peduncles green, it can also harm them by overexposure.

Temperature treatments

Temperatures and times of exposure are specified in the following recommendations. For many years, people have used heat treatments like hot water dips and hot air or vapour exposure to keep insects at bay (and for fungi, in some cases). Heat treatments were reassessed as quarantine treatments for fruits and vegetables such as mango, papaya, citrus, bananas, carambola and pepper, eggplant, tomato, cucumber, and zucchini when limits were extended to bromine-based fumigants. Commodity-specific temperature, exposure, and application methods must be followed carefully in order to avoid heat injury, especially in highly perishable crops. Once the treatment is complete, it is critical that temperatures be lowered to the appropriate levels for storage and transit.

Immersion in hot water demands fruit pulp temperatures between 43 and 46.7°C for between 35 and 90 minutes. This varies depending on the product, the insect to be controlled, and its development stage. Plums, peaches, papayas, cantaloupes, sweet potatoes, and tomatoes can all benefit from dipping in hot water, although bug control is not always guaranteed. Mangoes from Brazil should be dipped for 70-90 minutes at a temperature of 46.1°C and a depth of 12 centimeters. A pulp temperature of 40-50°C (up to 8 hours) or water vapour is required to kill insects in many tropical crops. Mango, grapefruit, Navel oranges, carambola, persimmon and papaya are all tolerant of hot air. Grapefruits, oranges, mangos, peppers, eggplant, papaya, pineapple and tomatoes have all been given USDA-APHIS approval for vapour treatment in the same way.

Sprout suppression

Sprouting and root production in potatoes, garlic, onions, and other crops hasten degradation. They also have an impact on the products' viability on the market. Sprouting or rooting products, on the other hand, are fiercely rebuffed by consumers.

During the "rest" period, bulbs, tubers, and some root crops finish their development. Reduced physiological activity and a lack of reaction to environmental cues are the hallmarks of this syndrome. If you put them in the right circumstances of temperature and humidity, they won't grow. Endogenous inhibitors like abscisic acid predominate over promoters like gibberellins, auxins and others when the plant is at rest. This equilibrium shifts to a "dormant" state as storage time passes. If given the right conditions, they'll sprout or develop roots. It's impossible to tell where one stage ends and the next begin. Slov: instead, the balance of pro- and anti-promoter molecules shifts. Promoters predominate and sprouting occurs as a result of longer storage durations. Chemical inhibition is favored over refrigeration and controlled atmospheres because of their lower sprouting and rooting rates. Potatoes are treated with CIPC (tri-chloroisopropyl Nphenylcarbamate) prior to storage, while onions and garlic are treated with Maleic Hydrazide, which is sprayed on prior to harvest. Once the curing process has been completed, CIPC must be applied.

Gas treatments before storage

Grapefruits, Clementine's, avocados, nectarines, peaches, broccoli and berries all benefit from exposure to a carbon dioxide-rich atmosphere (10-40 percent up to a week) before storage. Insects can be controlled with higher quantities of pesticide (60-100%). It's not clear how this gas affects the human body. Inhibition of metabolism and ethylene activity is well-documented with long-lasting effects following treatment. The spore germination and growth of decay organisms is also hampered at greater concentrations (> 20%). Low oxygen levels (1%) also help preserve the quality and control insects in a variety of fruits and vegetables such as oranges nectarines papaya apples sweet potatoes cherries and peaches.

Packaging

For the most part, packaging is used to ensure that the product is contained in a container together with packing materials to prevent movement and to cushion the product (plastic or moldable pulp trays, inserts, cushioning pads, etc.) and to protect it from damage (plastic films, waxed liners, etc.). It must meet three primary goals. These are addressed to the following people:

» By standardizing the number of units or the weight of the package, you'll be able to more easily handle and promote your goods.

» Protection from physical harm (such as impact or abrasion), as well as environmental harm (such as temperature and relative humidity), is essential for products in all stages of transportation, storage, and distribution.

Assist purchasers by providing them with details such as the type of product, weight, number of units, selection or grade of the quality, name of the producer, country of origin, and so on. Nutritional value, bar codes, and other traceability information are regularly included in recipes. If the product is to be properly treated, a well-designed package must reflect that in its design. Packaging must be capable of withstanding moist conditions, such as hydro-cooling or ice-cooling; if the product has a high respiratory rate, it must have large enough apertures for gas exchange;

if the item dehydrates fast, the packaging must be good at preventing water loss; and so on. Special ambiances can be created inside packages using semi-permeable materials. This aids in preserving the freshness of the vegetables.

Categories of Packaging

There are three types of packaging:

» Consumer units or prepackaging.

» Transport packaging.

» Unit load packaging or pallets

This is known as prepackaging or consumer unit when the weighed product is delivered to the consumer in the same container it was made in. A typical example of this would be the amount of food a family consumes over the course of a month (300 g to 1.5 Kg depending of product). Standard materials include shrinkable plastic sheets over moulded pulp or expanded polystyrene trays, plastic or paper bags, clamshells and moulded PVC trays. Mesh bags containing 3-5 Kg of various vegetables, such as onions, potatoes and sweet potatoes are commonly sold. Packaging materials' colours, forms, and textures have an impact on their visual appeal. Fiber board or timber boxes weighing between 5 and 20 Kg are commonly used for transport or packaging for marketing purposes. The following conditions must be met: Biodegradable, non-contaminating and recyclable materials should be used in the construction of containers that are easy to handle, stack, and carry. Packages created for repeated usage should be able to endure the weight and handling circumstances they were designed for, and be able to fulfill the weight criteria or count without overfilling, so that the volume on the return journey can be greatly reduced.

Packaging materials that function as dividers and hold the fruit in place are popular in these types of packages. Inserts placed vertically, for example, are an option. Produce is separated into layers in trays, which accomplishes the same goal. Apples, peaches, nectarines, and a variety of other fruits are all known to contain them. Large fruits like watermelons, mangos, papayas, etc. are protected by individual plastic foam nets. Paper or wood wool, papers, or other loose fill materials can also be used.

Fruits and vegetables are still packaged in natural fibre containers in many impoverished countries. They're cheap, but they can't be sterilized or cleaned. As a result, they pose a risk of microbial contamination when reused. It's also possible to bruise from compression. This is due to the fact that they were never intended to be stacked. In addition, the wide range of weights and volumes makes marketing a difficult endeavor.

Finally, pallets have taken over as the primary unit of packing both domestically and internationally. Containers, vehicles, forklifts, and other warehouse equipment all use the same dimensions. They reduce handling at every stage of the distribution chain because they are unit loads. Sizing options abound. However, 120 x 100 cm is the most frequent international size. Plastic is sometimes used in its construction. A pallet can store anything from 20 to 100 pieces, depending on the packaging size. Plastic tension netting or a combination of corner post protectors and horizontal and vertical plastic strapping is used to secure pallet loads. Sliding can be prevented by gluing individual packages together with glue that has a low tensile strength. They are also linked or piled crosswise to help stabilize the load. Sizes are becoming more and more standard. As a result, there are a wide range of packaging options for fruits and vegetables. One of the primary goals of standardization is to increase pallet surface use through the use of pallets with a common size of 120 x 100 cm. The ISO (International Organization for Standardization) module uses 40 x 30 cm and 30 x 20 cm components to partition the main horizontal dimensions of 60 and 40 cm. Individual packages' height is completely unrestricted. Palletized loads should not, however, exceed 2.05 m in height in order to be handled safely. The usage of non-returnable containers is a major environmental problem. There must be little waste of materials and recycling after the functional use of the package in order to limit the environmental impact.

Chapter - 10

Prevention of
Post-harvest Losses

In order to grow food, a farmer needs to invest both time and money, and unless he is doing so solely for the benefit of his family, he is forced to join the market economy, where he must recover his costs and turn a profit. More than two-fifths of food grains in the developing world are lost after harvest due to mismanagement and spoilage, as well as pest infestation; this means that one-quarter of what is grown never reaches its intended audience. A lack of proper harvesting, handling, and transportation of fruits, vegetables, and root crops can lead to their decomposition and unsuitability for human consumption. Sweet potatoes, plantains, tomatoes, bananas, and citrus fruits are among the crops that suffer the most in poor countries. Estimates of output losses in these countries are difficult to come by. This waste, especially if it can be prevented economically, is of enormous importance to both growers and consumers.

Causes of Post-Harvest Food Losses

As marketing processes become increasingly intricate, the factors that influence post-harvest food losses for perishables vary greatly. It's likely that a farmer raising fruit for his own family doesn't mind if his crop has a few scars and imperfections. In order to maximize his profits, he and his workers, if any, must adopt a different mindset if he is producing for a market outside of his neighborhood.

The grower may and must determine how significant look, maturity and flavour are to his product by knowing his market. In addition, the grower must determine if the packaging investment will actually pay for itself in enhanced crop value. Purchasing pricey containers for his produce will be pointless if the workers in the fields throw them around and damage what's inside. To reduce post-harvest losses, the grower must first adjust their own and their employees' attitudes, rather than relying on fancy packaging to instantly cure their problems and boost their profits. Attention must be paid to the following:

» Market demand for the products he will grow; he must know the market and his buyers.

» Cultivation.

» Harvesting and field handling.

» Packing or packaging.

» Transport.

» Market handling; possibly storage or refrigeration.

» Sales to consumers, whole sellers or agents.

» Perishability of the produce.

These and other issues will be addressed in the sections that follow. In the long term, it may be less expensive for the grower to make simple adjustments in attitudes toward post-harvest food loss prevention than it will be to make changes to the marketing chain's practices, such as containers or transportation upgrades. Among other things, he needs to teach his loved ones and field workers how to minimize his losses.

Contribution of Fresh Producet Human Nutrition

Foods from both plants and animals make up the majority of people's diets. The primary source of energy in the diets of the majority of people worldwide is starchy staple food, mainly cereal grains. Root and tuber crops, such as plantains and other related plants, are either the primary food source or a supplement to cereals in some regions, particularly in the wet tropics.

Eating a variety of fruits and vegetables is a good way to receive the vitamins and minerals your body needs. Some root (potato) and legume (pigeon pea, bean, and lentil) crops can provide a percentage of the protein needed as well as a wide range of flavour and colour options when consumed combined.

Energy requirement

» Energy foods are starches and sugars that the plant itself produces. Besides root and tuber crops, starch can also be found in plantains and green bananas.

» They're also a source of energy. Avocados, which contain between 15 and 25 percent oil, are the only fresh produce that contains significant levels of oil.

Food for body growth and repair

» To develop and repair muscles and organs, proteins are necessary. Growing children require a lot of these nutrients. On a dry-weight basis, some root crops like sweet potatoes and potatoes, as well as the leaves of numerous crops, contain protein values that are close to those of animal products. Fresh food, however, is low in protein content. Compared to other root vegetables, cassava has an extremely low protein level.

» When compared to caloric and protein-rich foods, minerals are necessary for good health, but their intake should not be excessive. Many trace elements, including as sodium, potassium, iron, calcium, and phosphorus, are also necessary. Calcium, iron, and a few other minerals are abundant in vegetables.

» When it comes to chemical interactions in the body, vitamins are vital. Vitamin C and other nutrients can be found in a wide variety of fruits and vegetables, as well as root crops. As seen in Table 1, fresh fruits and vegetables are a good source of numerous vitamins.

» Fresh produce contains a lot of "roughage," which is another name for fiber. Even though it cannot be digested, dietary fiber is critical to the digestive process, and medical research shows that a diet high in fiber lowers the risk of disease.

Loss of food value in fresh produce

Fresh produce's nutritional content can be affected in a variety of ways after harvest, including:

» With time, the dry-matter content decreases due to the depletion of stored food reserves by ongoing life processes within the product.

» After two or three days, the amount of vitamin C in the produce would have decreased significantly.

» Cooking degrades vitamins C. If they are grown and managed properly, raw fruits and vegetables can be extremely beneficial.

» If you're peeling potatoes, you'll lose a lot of their nutritional value because the protein content is just behind the skin.

The minerals and trace elements dissolved in the water used to prepare vegetables and fruit should not be discarded, but rather employed in soups or other culinary preparations.

Further information on the food value of fresh produce can usually be obtained at national nutritional councils or departments of health.

Pre-Harvest Factors in Produce Marketing

After harvest, fresh produce's freshness and condition cannot be enhanced. When and what a farmer plants, and how and when he cultivates and harvests his crop, all affect the eventual market value of what he produces. In general, farmers rely on their own experience and local traditions when it comes to crop selection and cultivation practices, but they may need to be referred to agricultural extension officers or research and development specialists at their national department of agriculture or an equivalent if they want or need assistance.

Market Factors for the Produce

Farmers' decisions on which crops to cultivate are influenced by the following factors:

» Neighbors, town residents, retailers, wholesalers, or intermediaries (commission agents) are all possible buyers of the produce.

» Size, shape, maturity, and appearance of the product.

» The buyer's price restrictions.

"Too excellent" and "too awful" commodities exist. A commodity that exceeds market needs may not bring in higher pricing and hence be a waste of resources.

Only a few variants of a commodity may be exchanged in most markets, which is a major drawback. The Agricultural Seed Experiment Station in East Java in Indonesia, for example, has recorded 242 mango varieties, but just seven have any commercial viability outside of select communities. Only by removing non-marketable mango trees and replacing them with desired ones can the local producer effectively improve his market share in the mango market. This variety criterion is crucial in international trade. Exporting countries have little choice but to provide what the importing countries will pay for. For the most part, this is true in the developing world. While the Association of South East Asian Nations (ASEAN) has promoted commerce in fruits and vegetables, many of which are native to the region, there are still regional preferences for specific cultivars.

It is difficult to establish new types as economic crops in poor countries. Traditional human conservatism may also be a challenge, unless there are significant economic incentives to break the cycle.

Influence of Production Practices

Pre-harvest production practices can have a significant impact on the quality and quantity of post-harvest returns and lead to the rejection or downgrading of produce at the time of sale. They include:

Water supply (Irrigation)

Transpiration and photosynthesis, the two processes by which growing plants transform light energy into chemical energy, both require a constant supply of water (the giving off by a plant of vapour containing waste products). bad outcomes may be caused by:

> » Green vegetables can become brittle and susceptible to deterioration if there is too much rain or irrigation.

> » Citrus fruits might have poor juice content and thick peel due to a lack of rain or irrigation.

> » Potatoes and tomatoes can develop growth fissures or secondary growth if they are exposed to dry circumstances followed by rain or watering.

Soil fertility, use of fertilizers

The quality of fresh produce during harvest can be adversely harmed if the soil does not contain sufficient plant foods. The development and post-harvest condition of the produce can be harmed by using too much fertilizer, though. Some of these outcomes include:

> » The yellow-red browning of leaves of green vegetables, such as cabbage, can be caused by a shortage of nitrogen.

»   Poor fruit development and irregular ripening can be caused by a shortage of potash.

»   Tomato blossom end rot and apple bitter pit are both symptoms of calcium-moisture imbalance.

»   In papaya, sunken stems, and beet peel breaking can all be symptoms of a boron deficit.

Soil-nutrition issues like these can be easily discovered during harvest. Temperature, moisture, acidity, as well as interactions between different fertilizer compounds, are all factors that affect the balance of nutrients in soils and their influence on crops. If you have a serious soil-nutrition problem, you should seek out the guidance of a soil-nutrition expert.

Cultivation practices: Crop management is critical to getting high yields and high-quality fresh fruit. Among the most essential aspects are:

»   Agricultural diseases are sometimes transmitted by weeds, and weeds in fallow ground near crops are just as essential as those in the crop itself. In addition to competing with crops for nutrients and water, weeds compete with each other.

»   The post-harvest deterioration is brought on by rotting plant leftovers, dead wood, and decayed or mummified fruit. In order to reduce post-harvest losses, they must be collected and removed.

Agricultural chemicals

These are two types:

Herbicides and pesticides are applied to the soil or sprayed on weeds, illness, and insects as a preventative measure. Improper use can cause damage to produce, resulting in spray burns, and improper disposal leaves harmful residues on harvested produce. Pesticide use is regulated in most nations, and only at the recommended concentrations. To avoid hazardous spray residues reaching the customer, it is imperative that the recommended time between the last spraying and harvesting be strictly adhered to. Extension or other agriculture department staff should provide advice on regulations.

In order to increase the marketability of fruit, growth-regulating chemicals are employed in the field to limit fruit set and promote uniform ripening. In the context of small-scale manufacturing, they are of limited significance. They are best suited for large-scale commercial manufacturing, and their use demands specialized understanding. When it comes to fruit and vegetable growers, harvesting a crop is a crucial decision point. In most cases, fresh produce is ready to be harvested when it has reached its peak nutritional value. Harvest maturity is the term used to describe this state. When a plant reaches the stage of flowering and seed generation, it has completed its active growth and reached the term "maturity," which can cause some misunderstanding. It is important to consider how long it will take to get to market and how it will be handled en route when determining when the "fruit" is ready for harvest. Due to a delay in harvesting, a lot of crops are picked early.

Most growers decide when to harvest by looking and sampling. Judgments are based on:

» Sight-colour, size and shape.

» Touch-texture, hardness or softness.

» Smell-odour or aroma.

» Taste-sweetness, sourness, bitterness.

» Resonance-sound when tapped.

This form of evaluation relies heavily on personal experience. It may take some time for newcomers to understand the ins and outs of cultivating fresh vegetables. When the green tops of bulb onions fall off and the green tops of potatoes die off, they are ready for harvest. Avocados, on the other hand, remain green and immature even after harvesting.

» Commercial growers on a large scale use a combination of observation and more complex measurement.

» Time-recording from flowering to harvest.

» Environmental conditions, measuring accumulated heat units during the growth period.

» Physical properties including shape, size, specific gravity, weight, skin thickness, hardness etc.

» Chemical properties (important in fruit processing, less so in vegetables), sugar/acid ratio, soluble solids content, starch and oil content.

» Physiological characteristics including respiration rate, acidity or alkalinity (pH).

Harvesting decisions will be made based on current market worth of the predicted yield, and also how long it will be before it is no longer a viable crop. Harvesting too early or too late to take advantage of better pricing at the start or end of the season is common practice for growers of seasonal crops.

Perishability and product losses

It is important to remember that all fruits, vegetables, and root crops are live plant components that contain between 65 and 95 percent water. The length of their post-harvest existence is largely determined by how quickly they consume the food and water they have accumulated. When food and water supplies are depleted, the produce dies and decomposes. Anything that speeds up this process has the potential to render the product unusable before it has a chance to be consumed. There are many factors that contribute to the loss of fresh food, and these factors all interact and are affected by external circumstances such as temperature and relative humidity, which are addressed below.

Physiological deterioration

Temperature, lack of humidity, and physical injury all contribute to an increased rate of natural deterioration, which in turn causes an increase in the rate of loss. When fresh product is exposed to temperature extremes, atmospheric alteration, or contamination, abnormal physiological degradation ensues. A variety of undesirable effects, such as a lack of ripeness or flavor might arise as a result of this.

Mechanical damage

Because of bruising and skin fractures caused by careless handling of fresh fruits and vegetables, the rate of water loss and regular physiological breakdown is greatly increased. Bacteria can enter the body through skin wounds and cause deterioration.

Diseases and pests

All living things are vulnerable to parasites. Diseases present in the air, soil, and water can affect fresh food before or after harvest. While some infections can infect produce even if the skin is unbroken, an injury is necessary for others to take hold. Fresh food is frequently harmed in this way, and the resulting damage is most likely to blame.

The post-harvest operations outlined below have a significant impact on the influence of all three factors. These factors all have a significant impact on the product's marketability and the price that consumers are willing to pay for it.

Types of Fresh Produce

A wide range of plant parts from a wide range of plant groups and species are now available for sale on the fresh produce market. For horticultural and domestic purposes, the categories fruit, vegetables, and root crops have no genuine botanical significance, but are terms of convenience. It's easier to classify them as commodities according to their post-harvest handling and storage response as well as their edible plant part types.

Roots and tubers

These are sections of plants that are designed to store food. Plants rely on them as a source of food when the weather is unfavorable, and they help the plant grow quickly when conditions improve. These are a few examples:

Edible part

> » Swollen stem tuber: Irish or white potato.

> » Compressed stem tuber (corm): Dasheen, Root tuber (from fibrous root): Sweet potato.

> » Root tuber (from main tap-root): Carrot, turnip.

There are a few exceptions to this rule, such as the sugars found in carrots and other tap-root tubers.

Edible flower

Plant breeders have created a variety of vegetables with rich blossom heads that are edible as immature buds. For a while, these were popular in temperate nations, but in recent years, cultivars that can be grown in heated conditions or at greater altitudes have been produced in the tropical regions. An important tropical fruit, pineapple, is formed by the union of a large number of immature and unfertilized blooms surrounding the plant's main stem, which eventually becomes the fruit's core.

Vegetative growth (leaves, stems and shoots)

These common green vegetables are important sources of minerals, vitamins and fiber (roughage) in the diet.

Reproductive structures

Seed-bearing structures, consumed mostly for their flesh, but with fleshy portions. When ripe, these well-known fruits have high sugar content, thus they're typically consumed at that point in their life cycle. Tomatoes and peppers, for example, are commonly used in salads and other dishes. Several vegetables, such as the green seed pods of several crops, are consumed before the seeds have hardened.

Post-Harvest Physiology of Fresh Produce

By absorbing water and carbon dioxide gas through their roots, green plants employ sunlight to produce sugars by mixing these two elements. Photosynthesis is the name given to this process. Starch is formed by combining sugar units into long chains and storing them as they are in the plant. Carbohydrates, such as sugars and starches, are chemical compounds that plants store for later use as a source of energy for growth and reproduction. As root crops go into dormancy, they store starches in order to provide energy for new development when the dormancy phase ends. Respiration, which happens in all plant parts before and after harvest, releases the energy needed for growth in both circumstances.

One of the reasons for post-harvest losses in fresh food is the deterioration of the plant's physiology. Physiology is the study of living organisms' internal processes. All of these life processes are still going on when fresh produce is picked and prepared for consumption. Because the crop can no longer create new food or water, it must rely on its own stores, which deplete over time, causing the product to deteriorate and eventually perish. Produce that is not harmed or attacked by decay organisms will eventually become unfit for human consumption due to this natural rot. Respiration and transpiration are two of the most important physiological processes that contribute to ageing.

Respiration

Plants use respiration to get oxygen into their systems and get carbon dioxide out. Carbon dioxide and water are released as a byproduct of the breakdown of carbohydrates in the plant. Heat is generated as a byproduct of this process.

All plant material, in the field or after harvest, undergoes respiration as a basic reaction. As long as the plant's leaves are producing carbohydrates, it can't be stopped without causing damage to the plant or its harvest.

Carbohydrates and water in fresh produce are non-replaceable after harvest. Respiration depletes stored starch or sugar, and as a result, the produce begins to deteriorate and eventually dies.

Effect of air supply on respiration

A steady supply of oxygen is critical to sustaining life. In normal lant respiration, starch and sugars are broken down into carbon dioxide and water vapour by the action of the air. Fermentation happens instead of respiration when the air supply is constrained and the available oxygen in the environment drops to less than 2%. For example, fermentation breaks down sugars to ethanol and carbon dioxide, which generates disagreeable flavours in food and promotes premature ageing.

The effect of carbon dioxide on respiration

Carbon dioxide builds up around produce that isn't properly ventilated due to a lack of access to fresh air. Bad flavours, internal breakdown, failure to ripen fruit, and other aberrant physiological states can be caused by a concentration of this gas in the environment of between I and 5%. Produce must therefore be properly ventilated.

Transpiration or the loss of water

When a piece of fresh food is harvested, it contains anywhere from 65 to 95 percent water. Water is constantly circulating within the growing plants. The roots and stems take up liquid water from the soil, which is subsequently transported to the aerial parts of the plant, particularly the leaves, where it evaporates.

A plant's transpiration stream is the water that flows through it. High water content is maintained, and the plant is supported by the internal pressure. Plants will wilt and eventually die if they don't get enough water. All plant parts are covered in a waxy or corky coating of skin or bark that prevents water from evaporation. The plant loses water naturally solely through microscopic pores, the majority of which are found on the leaves. Plant surfaces have pores that can open and close in response to changing atmospheric conditions, allowing for a more or less controlled rate of water loss while still maintaining the firmness of the growing sections.

Unlike a growing plant, fresh food can no longer replenish its water supply from the soil after harvest, and therefore must rely on its own water reserves. After harvest, fresh food loses a lot of water, which causes it to shrink and lose weight.

As soon as the fruit or vegetable loses 5% to 10% of its fresh weight, it begins to wilt and is rendered useless. Water loss from produce must be minimized as much as possible in order to keep it fresh for longer.

The effect of moisture content of the air on water loss

All plants have internal air gaps that allow water and gases to flow in and out of all of their components. Water vapour from the transpiration stream and the water produced by respiration can be found in the air in these areas. As the plant's internal pressure rises, water vapour is forced out through the plant's pores. Because of the difference between the pressure of water vapour inside the plant and the air, the pace at which water is lost from plant parts varies. Fresh produce must be stored in a moist environment in order to minimize water loss.

The effect of air movement on water loss

The more quickly the surrounding air travels over fresh food, the more water is evaporated. The removal of respiration heat requires air flow through product, but the rate of movement must be kept as low as feasible. As possible. Controlled air flow through food can be achieved through the use of well-designed packaging materials and appropriate stacking patterns for crates and boxes.

The Influence of the type of produce on water loss

A wide range of fruits and vegetables have different water loss rates. Because of their thin, waxy coating and numerous pores, leafy green crops like spinach lose water quickly. There are some foods that are more water-resistant, such potatoes, which have a thick corky skin with few pores.

One of the most important contributing factors to water loss is the type of plant part's surface area to volume ratio. As the volume decreases, the amount of water that is lost through evaporation increases as well.

Ripening of Fruits

Ripening is a natural process that occurs in all fruits and vegetables. At this point, the fruit is said to be fully developed. The fruit begins to decompose as soon as it reaches the point of ripeness. Fruits including aborigine, sweet pepper, tomato, breadfruit, and avocado are included in this category.

Different types of fruit ripening have distinct respiratory processes.

The term "non-climacteric fruit ripening" refers to fruits that ripen while attached to their parent plant. Premature harvesting reduces the fruit's sugar and acid levels, which lowers its quality for consumption. The rate of respiration decreases as the plant grows and as it is harvested. When it comes to maturation and ripening, time is of the essence. Cherry, cucumber, grape, lemon, and pineapple are just a few examples.

When a climatic fruit ripens, it means that the fruit is ready to be harvested but has not yet begun to ripen. Natural or artificial ripening can be used for these fruits. Respiratory climacteric is a term used to describe the rapid increase in respiration rate that occurs at the beginning of ripening. As the fruit ripens and improves in taste and texture after the climacteric stage, its rate of respiration reduces. There are numerous fruits and vegetables that fall into this category.

Artificial ripening is used in commercial fruit production and marketing to regulate the rate of ripening, allowing for meticulous planning of transportation and distribution.

The effect of ethylene on post-harvest fresh produce

Most plant tissues produce ethylene gas, which is well-known for helping fruits get started on the ripening process. In the marketing of fresh product, ethylene is crucial because:

» Commercial climacteric fruit ripening is possible with this technology. Fruits from faraway markets can now be collected while still green and sent to local marketplaces where they can ripen under ideal conditions.

» In storage facilities, fruit's naturally occurring ethylene production can cause havoc. Even a small amount of the gas can have a devastating effect on flowers. Lettuce and other vegetables that are mature but not quite ripe might be destroyed by ethylene if stored near fruit that is ripening.

» When fruits are damaged or infected by moulds, ethylene production increases. During transport, climacteric fruit may begin to ripen, resulting in early ripening. In order to avoid rotting, all produce should be handled with care. Produce that has been damaged or is deteriorating should not be stored.

Tropical citrus fruit, even when completely mature, remains green on the tree. In order to reach its maximum colour potential after harvest, it must be "degreened" using ethylene gas. De-greening can only be economically viable for high-value export or domestic markets because the gas concentration, temperature, humidity, and ventilation can be properly controlled in specialized facilities. Locals in most tropical countries are fine with ripe green citrus fruit.

Post-Harvest Management of Fresh Produce

Fresh food can be physically damaged for a variety of reasons, the most prevalent of which are:

Mechanical injury:

Because of their high moisture content and fragile texture, fruits, vegetables, and root crops are vulnerable to mechanical damage at any point of the supply chain, from the farm to the store.

> »   The use of shoddy harvesting methods.

> »   Field or marketing containers with splintered wood, sharp edges, or improper nailing or stapling are all examples of unsuitable containers.

> »   Inadequate or excessive packing of marketing or field containers.

> »   Dropping or stomping on produce while grading, transporting or marketing it might lead to injury.

Injuries caused can take many fonns

> »   Splitting of fruits or roots and tubers from the impact when they are dropped; internal bruising, not visible externally, caused by impact.

> »   Superficial grazing or scratches affecting the skins and outer layer of cells; crushing of leafy vegetables and others of produce.

> »   Injuries cutting through or scraping away the outer skin of produce will: provide entry points for moulds and bacteria causing decay.

> »   Increase water loss from the damaged area.

> »   Cause an increase in respiration rate and thus heat production.

» Bruising injuries, which leave the skin intact and may not be visible externally cause: increased respiration rate and heat production.

» Internal discoloration because of damaged tissues.

» Off-flavours because of abnormal physiological reactions in damaged parts.

Injuries from temperature effects

When exposed to temperature extremes, fresh produce of any kind will be harmed. In terms of temperature tolerance, products can differ greatly. In terms of cold storage, their ability to withstand low temperatures is critical:

» At temperatures between 0 and -2 degrees Celsius, all produce is at risk of freezing. Frozen fruits and vegetables have a watery or glassy look. Temperatures of less than 32 degrees Fahrenheit should be avoided at all costs, as the shelf life of many products is limited. Fruit and vegetables that have thawed out of the freezer are more prone to rot.

» Chilling injury is a type of fresh produce damage that can occur at temperatures below freezing but not below freezing. Most of these crops are native to the tropics or subtropics, but they can also damage some temperate varieties.

As soon as the food is brought back to market (i.e., ambient) temperatures, symptoms of chilling harm may begin to manifest in the fruit and vegetables. Susceptible product must be stored at temperatures just above its LST when it must be preserved for an extended period of time. Consequently, the marketing life of sensitive crops will be reduced since respiration has continued at a faster pace than typical in cold storage at higher temperatures.

If fresh produce is exposed to high temperatures generated by solar radiation, it will quickly decompose and lose its nutritional value. Temperatures of up to 50 degrees Celsius can be reached if produce is left out in the sun after harvest. It will overheat and decompose if shipped or transported without refrigeration or proper ventilation. Water loss from root crops and leafy vegetables is exacerbated by prolonged exposure to the tropical sun.

Diseases and pests

Fresh vegetables are frequently ruined by fungal and bacterial diseases. Even though viruses can cause significant losses in growing crops, they are not a significant post-harvest issue to be concerned about. Post-harvest losses in fresh produce are rarely caused by insect pests that are primarily responsible for grains and grain legumes. Locally significant pests, such as the potato tuber moth, are common.

***Diseases*:** There are two main types of losses in fresh food due to post-harvest illness The more catastrophic loss in quantity occurs when the contaminated product is rendered unusable due to profound rot. Before harvesting, the produce may have been infected with a disease. When the disease merely affects the surface of the produce, quality suffers. Commercial crops may lose their value as a result of skin imperfections caused by this pesticide. Affected skin can often be removed and the intact inner can be used in crops cultivated for local consumption.

Microscopically small fungal and bacterial spores are the primary means by which these diseases are transmitted. They can be found in the air, soil, and decaying plant material. Infected produce can be spread:

Abrasions from casual handling, damage from insects or other animals, or growing cracks natural pores in the above- and below-ground sections of plants, which allow the flow of air, CO_2 and water vapour into and out of the plant; through the by means of the plant's undamaged skin being directly penetrated. The incubation period varies from crop to crop and illness to disease. Pre-harvest or post-harvest, it can happen at any time.

An infection in the fields before to harvest may not be detected until after the crop has been harvested. Soil moulds, for example, can trigger the degradation of root crops while they are in storage. To a similar extent, tropical fruits that become infected at any stage of development can only exhibit signs of rot as they mature.

At any point between harvest and consumption, a food product may become infected. Most of the time, moulds or bacteria have infested harvesting or handling injuries.

The use of infected seed or other planting material might spread post-harvest infections in the field before harvest. Weed plants and other crops can serve as alternate or alternative hosts for a variety of illnesses. Infected soil on farm equipment, trucks, boots, etc., as well as crop wastes or rejected produce rotting in the field are other ways they spread. Another way that post-harvest infections might be transmitted is by-

» Dirt or decomposing produce in the field boxes.
» Contaminated water that was previously used to clean produce.

» Packaging; rotting and decomposing rejected product.

» Packaging facilities; contaminating nutritious goods in packages.

Pests: However, a small number of fresh crop losses are due by attacks after harvest. Localized attacks by pests, such as insects or other animals, can be extremely damaging.

» Insect larvae tunneling through produce, such as the fruit fly, sweet potato weevil, and potato tuber moth are the most common source of insect damage. Before harvest, pests tend to infest. Post-harvest spread is an issue when produce is stored or transported for a long period of time.

» Rats, mice, and other farmyard pests are now again a concern while storing produce on the farm, as they were previously.

Loss sssessment: After harvest loss is difficult to measure because there are no standard procedures. Whatever method of evaluation is employed, the outcome can only be applied to the given situation.

Accurately measuring losses in a current marketing organisation might be difficult. Losses may be assumed to be excessive, yet there may be no data to back up this assumption because of the following:

» No records are available.

» If records exist, they don't go back far enough.

» This data is only anecdotal, based on the observations of a few different people.

» If losses were only assessed when they were extraordinarily large or low, records may not accurately reflect the current state of affairs.

» For commercial or other purposes, loss estimates may be intentionally over- or understated to earn benefits or avoid disgrace.

Since accurate records of losses have not been kept over a period of time, a reliable assessment of the potential cost-effectiveness of ways to enhance handling practices is almost impossible, and the grower's marketing position is difficult to strengthen. It is obvious that the producer who wants to minimize post-harvest losses must keep accurate data.

Harvesting and Field Handling

Care given during harvesting and field handling directly affects both the quality and condition of the produce shipped to market and its subsequent selling price. The design and implementation of harvesting operations must adhere to a set of fundamental principles, regardless of the size of the operation or the resources of labour and equipment available. The grower's goal should be to:

» The act of gathering a good crop in good shape.

» Keeping the crop in good shape until it is consumed or sold.

» To sell the produce as soon as it is harvested to a buyer or through a market.

Planning

Successful harvesting and marketing requires careful planning from the beginning of the production process, especially in reference to:

» Choosing the right crop at the right time to suit anticipated market demands.

» Connections with buyers so that when the crop is ready for harvest, it may be sold for a fair price.

» Organizing labour, equipment and transportation in advance of harvest activities.

» Offering complete oversight throughout the entire harvesting and field handling process.

Labour

The labour supply won't likely be an issue with small-scale family production for regional markets. More demanding standards will need to be satisfied in terms of training and supervising labour as commercial production size and the distances between rural producers and urban consumers grow. Spending more money on the correct handling and packing of the fruit before it leaves the farm is economically sensible in terms of return. Growers will be required to train their own field employees while accepting any assistance that local extension personnel may be able to offer.

Training workers

All employees should receive general produce handling instruction, and those doing activities requiring a higher level of ability should receive more specialized instruction. General instruction everyone involved in harvesting and field handling should get the following general training:

Demonstrations of damage to produce, highlighting the necessity for constant caution in handling to prevent mechanical injury from things like:

» Wood packaging with sharp edges, splinters, or protruding nails or staples.

» Cramming too much into containers that will be stacked.

» Causing damage to food with long nails or jewellery.

» Distantly dropping or hurling objects into containers.

» Disposing of field containers in a rough or careless manner.

An explanation of why it's important to prevent crop contamination from things like:

» Laying the produce straight on the ground, particularly moist ground.

» Utilizing soiled harvesting or field containers that have been polluted with soil, agricultural residues, or rotting produce: containers must be maintained clean.

» Direct contact with chemicals other than those used for authorised post-harvest treatments, such as oil, gasoline, or any other substance.

Specific training

Workers entrusted with specialized duties such crop selection and harvesting, as well as post-harvest crop selection, grading, and packing (if necessary), should get specialized training. The following will be illustrated and explained in this:

» The procedures for determining when a crop is ready to be harvested and for discarding unwanted output in accordance with market demands.

» The specific harvesting method to be used, such as breaking stems or plucking, clipping, cutting, or digging.

» Using harvest containers and moving food into marketing or field containers.

» Choosing marketable produce at the point of field assembly, and, if necessary, grading it for size and quality.

» The proper use of post-harvest treatment (in which product is to be packed on the farm directly into marketing packets), such as wax coating or fungicides.

» The procedure used to pack containers or market products.

When the crop is prepared for harvest, labour, transportation, and operations are planned, the choice of when to begin harvesting will mostly depend on:

» Climatic circumstances.

» The market's condition.

Depending on the crops, the marketing date may be flexible. Some, like root crops, can be grown, harvested, and sold over an extended period of time or held on the farm while waiting for a good price. Others, like soft berry fruits, need to be sold right away or they would go bad. The ideal time of day must be taken into account once the decision to harvest has been made. The goal is to deliver the product to the market in the best condition possible, which includes keeping it as cool as possible, carefully packing it, and keeping it damage-free. The basic guidelines to follow are:

» Early morning or late afternoon is the coolest time of day to harvest.

» Avoid picking produce that has been damp from dew or rain. If not properly aired, wet produce will overheat and degrade more quickly. When produce is damp, it may be more susceptible to harm, such as rind breakdown and oil staining in some citrus fruits.

» When conveyance is not immediately accessible, safeguard collected vegetables by placing it under open-sided shade in the field. Produce that is left in the sun will become very warm.

» Early in the morning is a good time to pick food for local markets. If sufficient transportation can be obtained, it may be advantageous to harvest in the late afternoon and transfer to markets the following morning.

Harvesting Techniques

By Hand: The majority of produce collected for domestic rural and urban markets in developing nations is done so by hand. A certain level of mechanization may be advantageous to larger commercial producers. But the majority of the time, only agro-industrial production of cash crops for processing or export or both will require sophisticated harvesting equipment. In most cases, hand harvesting, when done correctly, will result in less produce damage than machine harvesting.

Hand harvesting is typical when fruit or other produce within the crop is at different stages of maturity, i.e. when multiple visits are required to harvest the crop throughout time. Only when a crop is harvested in its entirety at once can machines be used to harvest it.

Root and tuber crops

The majority of staple roots and tubers that grow underground are susceptible to mechanical damage during harvest due to digging implements, such as wooden sticks, machetes (or cutlasses, pangas or bolos), hoes, or forks. These crops can be harvested more easily if they are raised in mounds or raised beds, as is typical when producing potatoes. This makes it possible to push the digging instrument beneath the roots or tubers, which can then be levered upward to remove the soil and reduce the likelihood of crop damage. Similar techniques can be used to free other root crops from the ground, including taro, carrots, turnips, radishes, etc. by angling the instrument into the ground and pulling the roots upward. If celery has been earthed up or buried to blanch the stems, this technique can also be applied to that vegetable.

Vegetables

Vegetative growth can be harvested entirely by hand or with sharp blades, or in part. Knives must always be kept sharp and clean to prevent the transmission of viruses that cause diseases from plant to plant. Different plant components require different harvesting techniques:

»   Lateral buds (Brussels sprouts) and leaves only (spinach, rapeseed, etc.): the stem is manually snapped off.

»   The plant's above-ground portion (cabbage, lettuce): trimming is done in the field by cutting through the main stem with a hefty knife (the cut stem must not be placed on the soil).

»   Bulbs (green onions, leeks, mature bulb onions): mature bulb onions, leeks, and garlic are loosening by using a digging fork as for root crops (such as carrots), while immature green onions are typically able to be removed from the soil by hand. For bringing bulbs to the surface and undermining them, basic tractor equipment is provided.

Flower structures

Broccoli can be snapped off by hand and then trimmed; mature flower heads (such as those of cauliflower and broccoli) can be cut with a sharp knife and trimmed in the field.

Mature flowers (squash, chayote, pumpkin): Flowers are plucked individually by hand or whole shoot-bearing flowers are harvested as a vegetable.

Fruits: The fruit stalks of many ripe fruits and some immature seed-bearing structures, such as legume pods, naturally break at harvest time. Unripe or immature fruits and seed-bearing structures are more difficult to remove off the plant without damaging the fruit or the plant itself. Cutting them from the plant with clippers, secateurs, or a sharp knife is the most efficient way to collect them. Long poles with a bag attached to catch the fruit can be used with the clippers mounted on them for tree fruits.

Plucking methods vary according to the kind of produce being harvested:

» For fruits with a natural break-point, such as an apple, passion fruit, or tomato, the best way to extract the fruit is to lift, twist, and pull.

» For mature fruits with woody stalks that break at the fruit and stalk junction, it's best to cut off the tree, leaving only centimeters' worth of fruit stalk attached to the tree. Mango, citrus, and avocado stem end rot can be caused by disease entering the scar on the stem.

» When harvesting immature fruits with fleshy stems such as zucchini, okra and papaya using a sharp knife is preferable than breaking the stem by hand, the rough break is more vulnerable to decay than a clean cut.

Mechanical aids

Due to the limited resources available to small-scale farmers in developing nations, mechanized harvesting techniques for "once over" crops are unlikely to be common. Mechanical aids can be used in small-scale commercial operations, especially if tractors are nearby. Jobs that may benefit from the usage of such aids include:

» It is possible to use tractor-driven harvesters to harvest root crops such as potatoes, onions, and maybe some other root crops.

» Produce from harvesting sites can be transported to a "waiting area" for transportation by tractor-drawn containers, pallets, and bins.

Harvesting and Field Containers

Direct packaging into marketing packages of harvested fruit eliminates damage caused by multiple handling and is increasingly employed by commercial growers. The technique is uncommon in rural areas, where fruit is delivered to nearby markets and expensive packaging isn't necessary. However for commercial producers, the packaging can be beneficial in ensuring that their product arrives at the market in better condition and at a higher price. Harvesting and handling should be done in a way that protects the product and prevents it from overheating.

Selecting field containers for harvesting

The harvester must be able to carry them easily as they move through the field:

» Citrus and avocados, which have firm skins, can be harvested using harvesting bags with shoulder or waist slings. They are lightweight and convenient to carry. Produce bags should have an aperture at the bottom for easy emptying into field containers; they should not tip over.

» Fruits that can be crushed, like tomatoes, are best harvested in plastic buckets or other containers. Ideally, the containers should have no sharp edges or protrusions that could harm the produce inside.

» The sharp edges or splinters of baskets used for harvesting might harm produce.

» Especially if they are large, boxes that aren't well-built may tip over and crush or otherwise damage their contents if they aren't properly supported.

» Commodity growers utilize bulk bins with a capacity of 250 to 500 kg when sending crops like apples or cabbages to large-scale packing facilities for sorting, grading, and packing. A forklift attachment on a tractor can be used to carry produce from harvesting locations to assembly centers.

Unventilated bulk bins should only be utilized for short periods of time, and produce should be shielded from the sun or rain if they are to be used in the field. Keeping produce in bulk for a lengthy period of time will cause it to overheat and rot, making it more susceptible to spoilage. Bulk bins transported over long distances need to be perforated to minimize the accumulation of heat in the contents of the containers.

Post-harvest Handling

Field and farm transport: Preparation for the movement of food within a farm field should begin prior to planting crops. Inappropriate vehicles can cause significant harm to farm products when transported over rough roads. Loading and stacking containers on trucks should be done with care so that their contents are not damaged if they shift or collapse during transport. Shock absorbers and low-pressure tires are essential for safe driving. Even at modest truck speeds, the jarring of heavy containers can wreak havoc on fruit.

Transport from the farm: The destination of the produce leaving the farm will usually be one of the following:

» Products are transported in small containers, sometimes by animals or animal-drawn carts, but the majority of the time by vehicles owned or contracted by producers; public transportation is occasionally utilized. Local farmers' markets are also known as "farm stands."

» If a truck is parked outside for a long time before unloading, only the top portion of the load should be covered; grass or leaves are not recommended for this purpose because they restrict ventilation and may be a source of disease; complete encasing of the load with teepees is not recommended because it is too cumbersome.

» Only if the produce is put in marketing containers on the farm can it be considered a city market.

Damage Suffered by Packaged Produce

In order to transport huge quantities of market-ready fresh food, it is necessary to break it down into smaller units that can be carried by a single person. Containers with capacities ranging from 3 to 25 kg and dimensions up to 60 by 40 by 30 cm work well for this. Other bulky items, such as complete bunches of bananas, are shipped in 25 or 50 kg bags, depending on the commodity (e.g. potatoes). Loose or knotted bundles of leafy greens can be sold without being packed. Traditionally, produce is transported to marketplaces in baskets, bags, and tray fashion in the majority of underdeveloped countries. Low-cost and widely available materials like dried grass, palm fronds or bamboo are often used to make these products. They're fine for short-distance shipments of fresh produce, but they have a slew of drawbacks when transporting large, heavy cargoes over long distances.

In order to minimize waste and maximize the efficiency of transportation, bulk commercial quantities of food require superior packaging. The goal is to prevent damage to the product during handling, transportation, and storage, as well as to offer uniformly sized containers that can be conveniently handled and numbered. Frequent weighing and handling can be minimized by using standard-sized packaging that is easier to stack, handle, and load. Paper and paper goods, wood and wood products, and flexible and stiff polymers are used to make a wide range of container kinds. For each type, we need to think about the potential benefits, costs, and added value it can provide. It's usually a good idea to keep packaging costs down. Even though plastic crates were more expensive than traditional bamboo baskets, a Thai study found that they were still useful after 20 times the number of journeys. This means that plastic crates cost about one-quarter of the bamboo baskets per journey, while the bamboo baskets cost five times as much. In addition, the crate made it easier to transport vegetables, store it, and clean it. Small-scale growers may find it more cost-effective to use indigenous containers designed and constructed in-house rather than purchasing plastic crates.

Injuries

Cursor punctures

Cause: Splinters or staples poking through the packaging; sharp objects piercing the container; projecting nails or staples.

Effect: Produce with large holes or cuts, resulting in water loss and quick decomposition.

Impact (shock)

Cause: packages may be thrown or dropped; the car may be started or stopped suddenly, causing the load to move; or the vehicle may be sped up over a rough road.

Effect: bursting of packaging, bruising of contents Compression (squeezing or squashing)

Cause: containers that are too small or too large; containers that are either overfilled or stacked too high; or both. Transportation-related container collapse

Vibration (shaking)

Cause: vibration of the vehicle itself and from rough roads.

Effect: wooden boxes come apart, damaging produce.

From the environment

Heat damage

Cause: heat buildup in the produce due to insufficient ventilation in the package; and exposure to external heat, such as sunshine or storage near a heating system.

Effect: softens, wilts, and develops off-flavors in fruits and vegetables that are past their prime. A quick rate of deterioration dry and brittle cardboard cartons can be quickly destroyed by impact, making them more susceptible to breakage.

Chilling or freezing damage

Cause: temperatures below the chilling or freezing tolerance level during storage of delicate produce; low or subzero ambient temperatures.

Effect: Refrigeration-sensitive product is damaged; frozen produce decomposes when thawed; plastic containers become brittle and can break.

Moisture and free-water damage

Cause: condensation on packages and produce after they have been transported from a chilly store to a damp climate at room temperature; packaging wet produce in cardboard containers.

Effect: stacked cardboard containers become pliable and fall, squashing goods inside; damaged produce begins to deteriorate.

Damage from light

Cause: When exposed to direct sunshine, plastic bags and crates that have not been treated with a UV inhibitor break down.

Effect: When plastic sacks are transported, they break down, causing damage to the product. Plastic crates can be damaged or bruised by fracturing them.

From other causes

Chemical contamination

Cause: Damage to produce from preservative-treated containers, such as boxes made of wood treated with pentachlorphenate (PCP); contamination of produce from boxes impacted by mould growth; contamination of containers stored near chemicals.

Effect: Produce decay caused by contaminated moulds or flavour contamination, as well as surface damage and discoloration, can occur when produce comes into touch with the container. Boxes fall apart due to wood-rotting moulds.

Insect damage

Cause: insects present in packed produce; wood-boring insects in wooden boxes;

Effect: bugs like spiders and cockroaches can cause consumer and legal opposition, as well as the spread of wood-destroying insects in infected crates.

Human and animal damage

Cause: contamination and eating by rodents and birds; pilfering by humans.

Effect: rejection of damaged produce by buyers or inspectors; loss of income through loss of produce.

Cost-effectiveness Packaging

The price of the marketed product must account for the capital expenditure, unit packaging cost, and predicted profit associated with the use of packaging. It's difficult to estimate the value of packing because of a variety of elements, such as:

» In order to dramatically reduce losses.

» A product's presentation and quality might provide it an advantage in the marketplace.

» In some cases, the produce's marketability may be extended.

However, packaging must not surpass the market's readiness to absorb the product's increased value, i.e. the additional cost.

Prevention of injuries to produce

» Damage to fresh fruit during marketing can be minimized with the use of appropriate packaging and handling techniques.

» Wooden boxes or cardboard cartons must be carefully erected in order to prevent product damage during handling and transportation; nails, staples, and splinters are always a threat in wooden boxes.

» Loose-fill packs, which are particularly vulnerable to vibration damage, should be packed individually to prevent products rubbing against one another during handling and transportation.

» In the case of bruising, the cause is overfilling or collapsing containers; the collapsible boxes may be due to a weak wall, a softening of cardboard walls due to moisture, or the inability to stack boxes in such a way that the side and end walls support those above.

» Before shipping low-value product like root or tuber crops in 25 or 50 kg sacks, care should be taken to prevent damage to the bags from being handled forcefully. To prevent damage to the bags, stack them in pallet loads or pallet boxes when shipping them to reduce handling.

Effect of packaging on other types damage

Heat, Chilling or freezing

Overall, packaging is not very effective at keeping out heat or cold, as it has weak insulating characteristics. Due to a lack of ventilation in the packing, the produce may be damaged by the heat generated by it. When transporting vegetables with high respiration rates, expanded polystyrene packaging with good insulating characteristics are utilized topped with ice.

Most impoverished countries are unable or unwilling to afford such packages because of their high costs and limited availability.

Moisture and free water damage: Rain and high humidity damage cardboard boxes because they become soggy and collapse when they become wet. When it comes to manufacturing, this issue can only be solved by either waxing the cardboard or covering it in moisture-resistant plastic. Fruit and vegetables stored in moist sacks or boxes can deteriorate faster.

Chemical contamination: Outside sources of chemical contamination will not be prevented by packaging. As a result, the containers themselves get contaminated and contribute to the spread of disease. Unassembled hardwood or cardboard boxes, including "knocked down" sacks, should not be kept with chemicals.

Selection of packaging for fresh produce

Packaging can be a significant investment in the marketing of product, thus careful attention is needed when selecting containers for commercial-scale marketing. In addition to offering a uniform package size, a container must meet the following specifications:

» It should be portable and take up less room when empty than when filled.

» For example, nesting plastic boxes, foldable cardboard boxes, fibre or paper or plastic sacks.

» If possible, it should be straightforward to build by hand or with a simple machine.

» For transportation and storage, it must be able to offer appropriate ventilation for the contents; it must also be able to meet market expectations.

» In order to load neatly and securely, its dimensions and design must match the available transportation.

» The cost-benefit ratio must be favourable in proportion to the market price of the commodity being used.

» If possible, it should be easily available from a variety of sources.

Size and shape of package

It is important that packages are of a size that can be readily handled and that is acceptable for the marketing system in which they are being used. Wooden boxes, in particular, necessitate a smaller size to meet these requirements. It is critical to consider the container's weight in relation to the produce it holds. In countries where shipping costs are calculated based on weight, bulky packaging can have a considerable impact on the final selling price of a product. To maximize capacity and stability while transporting a load, a package's shape must also be taken into consideration. The amount of product that can be stored in a round basket is significantly less than that which can be stored in a box of the same size. Compared to a rectangular box, a cylindrical basket holds just 78.5 percent of the volume.

The need for ventilation in package

It is important to maintain the contents of any product well ventilated in order to prevent heat and carbon dioxide buildup. Transport and storage of food products in containers necessitate adequate ventilation at all times. Each package must be ventilated, but stacks of parcels must also be able to move air freely. Only if packages are constructed to allow air to flow through each package and across the stack can a tight stack pattern be acceptable.

It is necessary to stack sacks and net bags so that air can circulate within the contents. The amount of air moving through the cargo also affects the efficiency of ventilation during transit.

Packaging Materials

Packaging for fresh produce is of several types:

Natural materials: Throughout the developing world, baskets and other traditional containers are manufactured from bamboo, rattan, straw, palm leaves, etc. Reusable containers mean lower prices for both the raw materials and labour that go into their construction. The drawbacks include:

> » If they are polluted by decay organisms, they are difficult to clean.

> » When piled for extended distances, their lack of stiffness causes them to bend out of shape.

> » Transit; their form makes it difficult for them to load;

> » When they're packed too tightly, they create damage due to pressure.

> » Splinters or sharp edges can cause cuts and puncture wounds.

Wood: Due to rising costs, sawn wood is less frequently employed in the construction of reusable boxes and crates. Lighter boxes and trays are constructed from veneers of varying thicknesses. Reusable and stackable, wooden boxes are ideal for shipping because of their rigidity and reusable nature. The drawbacks include:

> » Multiple uses necessitate time-consuming and laborious cleaning procedures.

> » They are cumbersome and expensive to ship.

> » A protective liner is needed to keep the contents safe from sharp edges, splinters, and protruding nails.

Cardboard (sometimes called fiber board): Solid or corrugated cardboard is used to construct containers. Boxes and cases are the terms used to describe containers that close with a fold over or telescopic (i.e. separate) top. Trays are used to describe shallower, open-topped containers. They come in a collapsed form (meaning they're flat) and the user has to put them together themselves. There are a variety of ways to put together and close boxes: stapling, tying, taping, or using interlocking tabs.

Printed publicity and information about contents, amounts, weights, and measurements can be easily printed on cardboard boxes, making them a convenient and inexpensive packaging option. They come in a variety of shapes, sizes, and strengths. The drawbacks include:

» If just used once, they could be an expensive ongoing expense (however the boxes can be readily collapsible when empty if several uses are intended).

» Careless handling and stacking can easily cause damage to them.

» If they are exposed to dampness, they will be severely damaged.

» These items are only cost-effective to purchase in bulk; ordering in smaller amounts can be excessively expensive.

Moulded plastics: Many countries' product is transported in reusable boxes made of high-density polyethylene. A wide range of options are available. When full, they can be stacked on top of each other, and when empty, they can nest in order to save room.

Disadvantages are

» They can only be made cheaply in big quantities, yet even then, they are pricey.

» Most underdeveloped countries have to import them, which adds to the expense and frequently necessitates foreign money.

» They can be used for a variety of other purposes (such as washtubs) and have a high rate of pilferage.

» To be used in a regular go-and-return service, they need to be well organised and tightly controlled.

» Unless coated with a UV inhibitor, they degrade rapidly when exposed to sunlight (particularly in tropical regions), which drives up the price.

Natural and synthetic fibers: Natural fibers such as jute or sisal can be used to make produce sacks or bags, or synthetic polypropylene or polyethylene can be used to make the sacks or bags. As the name suggests, "bags" are typically used to refer to containers with a capacity of less than 5 kg. Both woven and netted options are available. The typical carrying capacity of a net is 15 kg. Potatoes and onions, which are less likely to be destroyed during shipping, are commonly packaged in bags or sacks, although even these commodities require extra care during transport to avoid damage. The drawbacks include:

» They are fragile and can be damaged by handling.

» They might be difficult to handle due of their size; if the bags are dropped or tossed, the contents will be severely damaged.

» If they're finely woven, stacking them causes problems with ventilation.

» Stacks may be unstable and collapse due to their smooth texture, which makes it difficult to stack on pallets.

Paper or plastic film: Packing boxes are commonly lined with paper or plastic film to minimize water loss and prevent friction damage. As much as six layers of kraft (heavy-wrapping) paper can be used to make paper sacks. They can hold up to 25 kg and are typically used for low-value produce. In the field, a simple tool can be used to wrap wire ties around the top of the bag and secure it. This method is recommended for large-scale crop production.

Disadvantages are:

» It's possible to waterproof paper walls by using plastic film or foil, but the bags will still collect gases and vapour.

» Heat from stacked food can take a long time to dissipate, causing damage to fruits and leafy vegetables.

» If bags are mishandled, limited protection for the contents.

Because of their low cost, plastic-film bags or wraps are commonly employed in the marketing of fruits and vegetables, particularly in consumer-size packs. In many developing nations, huge polythene bags are commonly used to transport produce; however this practice is discouraged and should be phased out altogether. The drawbacks include:

» In the event of an accident, they provide little to no protection.

» When temperatures fluctuate, they generate an excessive accumulation of condensation, which eventually leads to decomposition. They hold water vapour and so reduce water loss from the contents.

» If the bags are exposed to sunshine, the heat builds up quickly.

» This combination of vapour and heat with sluggish gas exchange results in fast degradation.

» Plastic bags should not be used for transporting vegetables, even if they have ventilation holes included into the design.

» Under tropical temperatures, plastic-wrapped consumer packets should be avoided, with the possible exception of stores with refrigerated display cabinets.

Deciding on Packaging for Fresh Produce

Many aspects must be taken into account by the grower or packing house operator before making a final decision on the packaging to be used. Market participants, packaging suppliers, transportation providers, and post-harvest extension consultants should be consulted before a decision is made. Consider the following:

» A specific type of food.

» How much product is lost throughout the marketing process currently.

» The cost comparison between the current packaging and that of the better packaging;

» Based on study findings, the projected decrease in losses if packing is improved reduced losses are expected to result in an increase in revenue.

» Is there a pre-packaged option? The cost per unit of packages decreases significantly when they are purchased in bulk; custom packaging is expensive.

» How often will the new package be available.

» Is there sufficient storage and assembly space for the protection of packing materials prior to use.

» Is the market going to accept the new packaging?

There can be no economic justification for introducing new packaging if it does not enhance sales. Good product that is adequately packaged has an advantage over produce that is badly packaged, and the returns from the investment can offset the costs of the packaging process. As a result, the marketing value of good packaging can be justified.

Post-harvest losses of fresh food may not be eliminated or considerably reduced by improved packaging alone. When it comes to marketing, packaging is just one of many factors that need to be taken into consideration.

Packing House Care

Sorting and packing of fresh produce is required whether it is sold in markets or directly to consumers or agencies. To prepare goods for market, packing houses, which can range from a basic, on-farm thatched shed to an automated regional packaging plant handling vast tonnage of a single commercial crop, such citrus fruit or apples, are used.

The packing house, whether simple or complicated, provides a sheltered environment for the assembling, sorting, selection, and packaging of produce in an organized manner with little waste.

Packinghouse size and design will rely on the type and volume of produce, market requirements, local infrastructure, the expected life term, and the projected cost of the facility. Things to keep in mind as you're developing your strategy include:

» Action items to be completed.

» Where about of a good place.

» The structure's design and the materials at hand.

» A list of the necessary tools.

» Management.

Operations

Any or all of the following operations may be performed, depending on the product or crops being handled and the target market being served:

» Reception: off-loading, checking, recording; sorting.

» Special treatments, if required (cleaning or washing, fungicide spraying, selection, size-grading).

» Packing; post-packaging treatments, if required (fumigation, cooling, storage); assembly and dispatch.

The all-too-common state of confusion where product is received, sorted, cleaned, dipped in fungicide, packed and stacked for delivery needs to be avoided at all costs. Each delivery should be: in cases where the packing company receives supplies from many producers.

» Labeled to indicate where it came from and when it arrived.

» Made sure of the amount or weight of the product that was delivered.

» Quality-checked samples if necessary.

» A receipt is provided to the vendor as proof of payment.

» The reception is a should be organized so that produce moves through the packing operation in the order it is received: first in, first out.

Sorting: Impurities and extraneous matter (such as soil or stones from the plants) should be removed from the produce during a preliminary sorting process. All waste should be carried away from the packing house as soon as possible, or stored in bins that can be sealed and removed at a later time. As a result, any garbage that accumulates in or near the packing house will contaminate produce that is to be sold to customers.

Cleaning and washing: Hand-picking or sifting can be used to remove the soil and stones listed above. Washing, brushing, or wiping down some fruit is an option.

» Hand-polishing or machine-brushing produce, especially fruit, can eliminate mild soil contamination or dust. Damage to the fresh produce's skin will accelerate deterioration, so proceed with caution.

» Washing is necessary to remove latex stains from fruits and vegetables that were damaged during harvesting, such as mangoes and bananas. Keeping in mind that washing is only necessary when absolutely necessary, it's crucial to mention. Fungicides should always be used soon after washing produce.

» Washing should only be done with fresh, flowing water. To avoid rotting produce, never wash produce in recalculated or stagnant water. These water sources can soon become overrun by decay organisms.

» Fresh produce washing water cannot be treated with antibacterial agents because none are acceptable or effective. Organic matter in the water, such as plant debris, quickly deactivates hypochlorite's or chlorine gas added to washing water for commercial treatment of some products. However, hypochlorite's or chlorine gas cannot be recommended for small-scale washing operations because it is rapidly deactivated by organic material. As it is difficult to monitor and replace chlorine concentrations in the wash water, the disinfectant's effectiveness against deterioration is restricted.

» To prevent fungicide from being diluted below its effective concentration, produce that needs to be treated with a fungicide should be drained after washing. A single layer of washed produce should be laid out on raised racks of mesh or slats, in the shade but with sufficient ventilation, to aid in rapid drying when washing is not followed by fungicide treatment.

Fungicide treatment

Moulds and bacteria are two of the most common causes of fresh product loss during the marketing process. Both before and after harvest, product can be infected by direct penetration of the undamaged skin. As the fruit ripens, some pre-harvest illnesses may come to life and become more dangerous as the fruit deteriorates. Latent anthracnose infections can be found in bananas, avocados, and sweet peppers, to name just a few.

When crops like apples, bananas, and citrus fruit must be stored for an extended length of time or transported over great distances to distant markets, post-harvest fungicide spraying is common practice. Fungicides are often sprayed after the produce has been cleaned and drained, as previously described.

Post-harvest decay control fungicides are typically applied as wet table powders or emulsified solutions. This means that if the mixture is not constantly stirred during its application, they will settle out of suspension. Thus, if the suspension is not constantly agitated, the fungicide concentration sprayed to the crop will fall below its effective level.

In small-scale packing operations, fungicide can be applied by

Dipping: You can use wire-mesh baskets to dip multiple small pieces at once, and then drain and dry the produce in a shady, ventilated area after it has been treated by hand with an agitated fungicide suspension.

Spraying: After washing and drying the produce, use a hand-operated knapsack sprayer to saturate the vegetables completely and to the point of runoff.

Using a mechanical mixer for the fungicide may be necessary for larger spraying operations that require a mechanized spray or drenching system. A conveyor belt or roller conveyor may be used to move produce through the spray or drenching process.

Only large-scale businesses storing produce employ other application methods such smokes, dusts or vapour.

Quality selection and size grading

There may be further quality and size selections made prior to packaging even when the product has already been sorted on the farm or upon arrival at the packing house. The breadth of these operations is determined by the market: will purchasers be willing to pay a premium for them. No costs for high-quality food? The quality requirements of urban clients are much more stringent than those of rural customers. Using only the human eye and sizing rings or gauges, selection and grading are best done by hand in tiny packing houses.

Waxing: Small-scale packing does not require specialized equipment to apply wax or similar coatings to improve appearance and reduce water loss from produce.

Packaging: Filling marketing containers by hand in small-scale businesses is called packaging. To pack long-lasting foods like potatoes and apples in large packing plants, machines are employed. However, these machines are costly and unsuitable for packing small quantities of varied products. There are a variety of ways to package a product:

» Where size grading is unnecessary, weighing is required, and loose fill jumble packs are utilized.

» The multilayer design pack includes a variety of produce, such as oranges, apples, and pears, which are offered by the pound.

» Size-graded multilayer mechanical packing packs have separator trays between layers; they are sold per box and are used in mechanical packing.

» High-value produce can be packaged in single-layer packs with each piece individually wrapped in tissue or placed in a separator to keep it separate; these are sold by the box.

Special treatments after packing

However, post-packing treatments are done to some crops in large-scale operations for urban and export markets, rather than in smaller operations. The most common methods include:

Fumigation: Pest management, such as fruit flies, is the primary goal of the treatment. There are several countries that need this, and it requires specialized equipment and highly-trained employees.

Initiation of fruit ripening: Ethylene gas is used to treat the fruit in insulated, temperature-controlled storage facilities for several days. The prices are prohibitive, limiting the application to large-scale projects.

De-greening of citrus fruit: When citrus fruits are ripe in the tropics they will remain green unless they are kept cool at night. When artificially degreened by ethylene treatment like that which initiates ripening, they will return to their natural colour. However, it is not typically done in small packaging houses.

Assembly of packed produce for dispatch: Delays in the marketing of fresh goods might result in significant losses. It's critical that produce is delivered to the market as quickly as possible after it's been packed. Consequently, the management of the packing house should place a high focus on transportation. Although it may take time to assemble a full load in small-scale operations, every effort must be taken to prevent the deterioration of packed goods that accumulates over time. The following points need to be taken into consideration: packed containers must be protected from the sun and rain; heat and water cause rapid deterioration of produce and seriously weaken card board boxes.

To prevent damage to the contents, carefully packed boxes must be managed during stacking. The loss of water and deterioration caused by harm to produce are both promoted. Because warming causes quick deterioration, containers awaiting shipment must be arranged such that they receive adequate ventilation.

Fresh produce losses can be minimized during packing procedures by:

» Keeping one's cool at all times.

» Protected from the elements by being kept dry.

» Safeguarded against harm.

» Continued to make rapid progress toward the market.

Planning a packing house: Consider the following when looking for a site for a packing house:

» The location of the planned market, the producing areas, and the transportation routes must all be considered.

» Is there a need for labour.

» Are utilities, such as electricity, water, and telephone, readily available.

Before deciding on a location, make sure the water used to wash products is free of contaminants such as sewage, factory effluents, pesticides, herbicides, and fertilizers by conducting a quality check.

Site characteristics: Following is a list of things to keep in mind when selecting a location:

» Sites should be flat and protected from severe winds if at all possible.

» If this is going to be a long-term packaging enterprise, the location needs to be expandable.

» Vehicles of all shapes and sizes are likely to visit the site; therefore there must be enough space for them to move around and park. At a minimum, all roads must be 3.5 meters wide.

» For rain runoff and packing water use, drainage needs to be adequate.

» Ideally, the site should be able to accommodate security measures like fencing, guards, etc.

Layout, Construction and Equipment

Packing operations on a small scale sometimes deal with a wide range of crops over an extended period. The layout of structures and equipment should be simple and adaptable in areas where volume is low.

Layout: Size matters when it comes to planning. Buildings with one end for receiving and the other end for dispatching are the most efficient layouts in most cases separating the reception area from the packing and shipping areas reduces the possibility of contaminated produce during the sorting and packaging process. It should also prevent traffic jams and confusion among arriving and exiting vehicles. However, this is not suggested due to the increased risk of contamination and congestion, not to mention the increased difficulty of future development that comes with a U-shaped plan that places reception and dispatch sections next to one other.

The area of the packing house should be adequate for the easy movement of produce through threes tags.

Reception: It is in this area that the product is unloaded, sorted and cleaned, including washed if that is necessary. Dirty with soil, dust, and rotting plant matter is likely to be found there in. To minimize contamination of the cleaned, sorted, and packed fruit, it is best if it is kept separate from the other procedures.

Preparation and packing

Drying equipment for washed or treated produce will be included in this area, as well as any specialized treatment options. There will be packing and grading facilities for the cleaned produce, if necessary. Additionally, there should be a dry area where packing supplies can be stored and assembled. As a whole, the area should be well-ventilated and well-lit to keep out the rain and snow. Keeping the selecting, grading, and packaging areas clean and dry is essential.

Dispatch: Next to the packing operation, this activity should be maintained fully separate from any permanent machinery. Temporary storage for packed produce must be large enough to allow for unrestricted movement of personnel and the goods they are transporting. There must be adequate airflow and sanitation in the dispatch area. It's likely that this is where any other offices or quality-control functions would be housed.

Construction: Materials and construction methods will be determined by the crops to be processed, the estimated volume of crops, the market to be served, and the funding available. As long as the structure is simple and economical, small-scale enterprises can succeed. The following are the primary requirements:

»   Appropriate protection from the sun and rain from the roof. In order to achieve this, the roof overhang should be at least one meter wide on all sides.

»   Adequate ventilation, but some protection from rain and dust from the wind. Overhanging roofs normally leave a large ventilation area behind the walls, which can be used to supply this.

Simple structures fashioned from inexpensive local materials (such as bamboo or bush poles, dried grass or other thatch) may suffice for small-scale packing on the farm. Although the lifespan of such a structure may be brief, the cost and ease with which it may be repaired or replaced more than make up for it. The interior of a building can be kept cold by periodically soaking dried plant material walls and roofs, if enough water is available. A corrugated sheet metal roof and walls on a concrete floor can provide a more durable tiny packing house. Sheet-metal structures generate severe heat in regions with lots of direct sunlight, hurting both the workers and the products they house. If you must use sheet metal, be sure to leave a large ventilation space between the walls and the roof, as well as a large overhang on the roof. In other cases, it may not be required to construct walls, especially if the roof is large enough to shield the product and the workers from the sun and rain. Non-slip concrete flooring placed with a fall to drainage channels for easier cleaning in permanent packing buildings is recommended. Concrete that has been treated with an anti-dusting surface coating has an advantage.

Fixed equipment should not be installed in packing facilities, unless they are used for large commercial operations. This gives you the greatest amount of flexibility when it comes to rearranging the arrangement based on the volume and diversity of crops.

Equipment: Depending on the size of the operation and the crops handled, each packing house will require a different set of equipment. Stuff'll be easy, and a lot of it can be created right here in town. Concrete washing tanks should be avoided since they are not moveable.

» Moving produce to the point of packing can be made easier by using bins or trays that can be carried by one person. They can be made from wood or plastic, but high-density polythene is the preferred material. There are advantages and disadvantages to both plastic and wood containers. It is possible to move many containers at once.

» Any two- or four-wheeled truck, such as those found in marketplaces or industries can be used as a push-cart.

» To transfer containers through several stages and to load and unload trucks where each container must be handled individually, roller conveyors supported on stands about 75 cm high are appropriate.

» Larger enterprises may benefit from the use of mechanized moving-belt conveyors, but these are more expensive.

» A mechanical hoist or a loading bay raised to the height of a truck bed is necessary if you want to employ a hand-pushed lift truck in a larger packing facility where mechanized means are needed to handle unit loads on pallets; these cannot, however, be used to lift full pallets onto vehicles.

» Palletized loads are moved around huge packing facilities using motorized forklift trucks.

When they're not in use (a), they nest to save space. To avoid crates nesting or crushing the contents, turn each crate in the opposite direction of that below to prevent them from stacking neatly and securely when full.

Selection, grading and packing: If there are any unmarketable items remaining after the initial sorting, they should be removed before the fruit is packed. The selection, grading, and packing of modest quantities of vegetables can be accomplished with a basic stand. In order to accommodate a higher volume of goods, the stand shown can be adapted to any desired length or even doubled if necessary.

Produce can be selected and graded by eye or by simple gauges, either hand-held or fixed, used by experienced employees. The packing bin is where graded and selected produce is deposited before being packed into shelf-ready containers. The dispatch assembly area receives the packed containers.

Additional equipment

Weighing: It's not uncommon for packing houses to need some kind of scale to weigh their produce, as so much is still sold and bought by weight. Before settling on a scale, it's a good idea to familiarize you with the many kinds on the market and the purposes for which they're most suited.

Washing: A galvanized tank can be used to wash produce in fresh running water. The water flowing from the entrance pipe, perforated on one side, over the end of the tank might propel floating produce along the tank. At the output end, a vertical baffle will help guarantee that all produce is thoroughly cleaned.

Drying: Before packing, produce that has been cleaned or treated with fungicide must be dried. A drying rack or table composed of wooden slats or plastic-covered wire mesh can be used in a small packing house to accomplish this. The drying table can easily be constructed from bamboo or bush poles if packing is done on the farm. Fungicide spraying can be done on the drying rack after washing, and the produce can then be left to dry before packing.

Floating produce will be pushed along the tank by water that enters through a perforated pipe. The baffle near the drain pipe aids in the movement of water through the vegetables.

Packing house Management

For packing facilities to function efficiently, managers need to be adept at coordinating the technical, organizational and economic components of their business.. Returns for growers will be negatively impacted in the event of any operational errors or delays. If it is financially feasible, operations should be maintained throughout the year.

Meeting the demands of the market - Workers in the fields and in the packinghouse should be able to get advice and instruction from management to ensure the most efficient operations and high-quality products.

Control and procurement: In order to run the packing house efficiently, it is critical to have accurate information about the size and arrival time of food. The packing house can arrange for the harvest to be picked up. Farmers that deliver their goods to a centralized packaging facility should be aware of the quality control procedures in place and the standards that are being upheld. Packing quality must also be monitored to prevent disagreements from occurring during marketing.

Packing supplies are available: Ahead of time, forecasts of the upcoming year's requirements must be made. As soon as possible, contact suppliers to secure the best possible rates and delivery dates. So that supplies do not run out during packing, accurate stock control must be performed.

Low-quality vegetables are thrown away: There will always be some inferior products in the process of market product selection and grading. The packing house may be able to get some use out of them, but it doesn't mean that they should be kept around for no reason. As a result, managers must also be aware of the amount of produce that is thrown away. It is necessary to account for the disposal of both low-quality and total-loss produce.

Staffing: The packing house's staffing must be sufficient for efficient operation, but labour expenses must be kept in mind. To put it another way, this means that workers must be efficiently deployed and that supervision must be adequate. An organization's full-time employees may comprise a management as well as clerks, mechanics, drivers, and possibly even some experienced packers. Temporary labour will be needed at busy times.

Staff development: In order to be in charge of all packing-house activities, the manager must be technically proficient and able to instruct his subordinates in their duties. In addition, he must ensure that the packing staff receives on-the-job training.

Training of the grower: A packing house's management should be aware of how to meet the quality criteria set by the market when it receives produce from multiple growers. Post-harvest extension staff should be included in the process. Formal training can take place outside of harvest season, but farm visits during harvest and packing-house operations are most successful.

Operations costing and accounting: Quality control criteria must be taken into consideration while negotiating with growers about payment. In order to minimize expenses and maximize profits for producers, it is necessary to evaluate the packing house's operating costs per kg of food processed.

Accounting and documentation: It is the manager's job to make sure those accurate records is kept and that operator accounts are properly managed. The packing house's overall success depends on this.

Prevention of Losses During Transport

Fresh produce marketing relies heavily on transportation, which is typically the most crucial component. In an ideal world, transportation would transfer produce directly from the farmer to the customer, as it does in many developing countries. Transport costs are often as high as or higher than the raw materials worth in more sophisticated marketing systems (such as those supplying towns, cities, or distant nations). It's possible to lose a lot of money because of bad transportation circumstances. Every person involved in transportation should strive to ensure that produce is transported in the best possible shape and that the transportation process is as efficient as feasible. Produce should be appropriately packaged and loaded onto a suitable vehicle in order to achieve this goal.

Causes of Loss

Mechanical damage and overheating are the primary causes of non-refrigerated transport damage and loss.

Mechanical damage

> » Careless packing during loading and unloading.

> » Vehicle vibration (shakes), particularly on bad roads.

» Fast driving; poor vehicle condition • poor stowage, which allows packages in transit to sway and collapse.

» Packages stacked too high; produce movement within a package increases in relation to its height in the vehicle are all causes of this damage type.

Over heating: Not only can heat from outside sources do this, but so can heat generated within the package itself by the product itself. Increasing the pace at which fresh fruit loses water is a side effect of overheating. A few of the more common reasons of excessive heat are as follows:

» Closed cars without any means of airflow.

» Patterns of close-stow stacking prevent air from moving between and through packages, limiting the passage of heat.

» An insufficient amount of airflow through the packages themselves.

» During shipment or while vehicles are waiting to be unloaded at their destination.

There are a number of strategies to limit the danger of food deterioration during transport. Produce trucks were common in the early days of the food industry. The majority of fresh food is now transported by road vehicles, with a small percentage travelling by air, sea, or inland rivers. Open pickups and larger trucks, both open and enclosed, are the most frequent modes of transportation. The following points should be kept in mind when driving on the road:

» The only exceptions to this rule are local deliveries from farmers or wholesalers to nearby retailers in closed vehicles without refrigeration.

» A roof can be attached to the frame of open-sided or half-boarded trucks. It is possible to use cal to close the open sides. Roll-up or sectioned vas curtains that can be used for loading and unloading at any location around the vehicle. The fruit can be protected from the outdoors while yet being able to breathe thanks to these types of curtains. The truck's sides and back must be covered in wire mesh if pilfering is a concern.

» White-painted roofs can be installed 8 or 10 cm above the main roof to reflect sunlight and keep produce cool; this can be done for a small additional cost.

» A more elaborate air intake can be used in conjunction with a louver to ensure a good flow of air through the load in long-distance vehicles.

» Long-distance transportation with refrigerated trucks, rail cars, or marine containers is an option, but it's prohibitively expensive for small businesses.

***Handling and stowage practices*:** When it comes to transporting fresh fruit, truck design and condition are critical considerations, but loading and stowing procedures in vehicles are equally crucial to prevent damage and loss.

» A stable and well-ventilated load must be attained in order to reach the best loading factor, which is the greatest load that can be carried economically under appropriate technical conditions.

» In order to ensure that the contents of the container are adequately ventilated, the package's size and design need to be carefully considered.

» The loading and unloading of vehicles must be closely monitored to avoid packages from being handled carelessly. Individual packages should be reduced to a minimum by utilizing loading devices such as trolleys, roller conveyors, pallet trucks, and forklifts whenever practical.

» It is imperative that stowage be done carefully to prevent the stow from collapsing during shipping, and goods should not be stacked higher than is allowed by the manufacturer.

» When loading and unloading, the packaged goods should be shielded from the sun and rain at all times.

» Loading on pallets or tonnage (wooden or slatted racks) will let air to circulate around the stacks as they are being transported.

» The order in which packages are loaded and emptied should be reversed if the load is to be transferred to multiple places; at the same time, the vehicle's weight should be evenly divided.

All the foregoing safeguards can be implemented, but the standards of driving remain an intractable problem. To earn more money for themselves or their employers, drivers are often encouraged to speed. Only qualified and trustworthy drivers should be engaged wherever possible.

***Other modes of transport*:** Many different methods are used to convey fresh produce, including head-loads and air-freight. The same rules should be applied in every situation. In order to qualify as produce.

Rail transport: In some countries a large amount of produce is carried by rail.

The advantages are:

>> When compared to the damage caused by haulage on difficult roads, moving transport damages to products are less.

>> Transport by air is more cost-effective than by road.

Because of the requirement for road transportation to get to and from the train station, rail transportation is more time-consuming and expensive than transportation by road alone.

Water transport

Inland: Some countries rely on waterways for the transportation of agricultural goods. Locally produced containers and bags are used to transport much of the food. As a result, there is no particular treatment offered for fresh fruit on the boats used.

Sea: In island nations, it is normal practice to carry fresh produce over short distances using tiny ships without refrigeration (e.g. the Philippines). As well as carrying passengers and normal cargo, ships may also store fresh food in unventilated holds. Due to poor packaging, hard handling by porters, and scorching in unventilated holds or near engine rooms, the losses are substantial.

This means of transportation has a lot of space for improvement. The refrigerated export of commercial commodities such as bananas is a good example of efficient and organized maritime transport, but a minimal investment by the small-scale shipper might substantially increase efficiency.

Freight by air: International trade in high-value exotic crops transported by air is well organized, much like shipping. Papua New Guinea, for example, has inadequate road connections, thus produce is flown in from rural areas to major markets. There are a number of factors that contribute to high costs and large losses: poor, non-standard packages;

>> Airports are notorious for their carelessness and exposure to the weather.

>> Passengers are given priority over cargo.

» Poor weather or mechanical problems might cause aircraft delays.

» Small produce shipments and short periods of refrigeration are among the conditions under which produce is stored.

It is unlikely that the general situation will improve much unless road links between producers and consumers are developed, even if adjustments are made to packaging and handling.

Special uses

Preparation of the fruit for packing includes cleaning, sorting, grading, and packing as standard procedures. In addition to these, commodities that are seasonal, exposed to long-term storage, or very perishable and carried over great distances to market require particular treatments to halt degradation and reduce losses. Temperature and moisture control are two of the most common methods for reducing fresh product losses, and these treatments can be done before, during or after packing.

Curing

Curing refers to the process of preparing starchy root vegetables like potatoes and onions for long-term storage. When it comes to root crops like potatoes and carrots, the cure process is very different.

Root crop curing

It is possible to cure root and tuber crops by replacing and strengthening regions of corky skin that have been damaged by water loss or decay organisms. The Irish potato is the most commonly treated crop, however curing can also benefit tropical root crops.

The following conditions must be observed regardless of the crop:

» New skin growth on the roots and tubers can only be stimulated by keeping them at a temperature that is slightly above ambient.

» Moisture must be maintained in the air, but not at the roots or tubers themselves, No new skin will develop on wounded surfaces if left to dry air.

» As new skin grows, it needs some ventilation, but an overabundance of it can dry the air and lower the temperature.

» In order to prevent bacterial soft rot, the temperature must to be kept constant. If it drops, water will condense on the roots and tubers.

As soon as possible after harvesting and handling, root and tuber crops should be cured to minimize damage. Allowing the temperature to increase enough to facilitate curing can be achieved by restricting airflow. Due to the roots' natural production of water and the rapid rate of evaporation from injuries, the air will become moist at the same time.

Only experimental data has been used to determine storage requirements for tropical root crops such as sweet potatoes and parsnips. Since sweet potatoes and other aroids like taro and cocoyam are highly susceptible to post-harvest deterioration, their storage life is usually quite limited. When left untreated, cassava quickly turns brown and rots from the inside out.

Curing dry bulb onions: Curing dry bulb onions is a drying procedure that begins as soon as they are harvested. An onion field must be kept dry and warm for a few days after harvesting before it can be stored for long periods of time. Drying onions on racks or trays in a covered area may be necessary if the weather is moist. Onion curing is required because:

» Onions' necks are particularly vulnerable to deterioration if they remain damp, especially if they've been exposed to sunlight for long periods of time.

» Before harvest, the green tops are removed.

» Drying the outer skins of the bulbs lowers decay and water loss, and broken roots are a common entry site for decay unless they are quickly dried.

A warm and moist environment is important to aid in the healing of skin injury when this method is correctly implemented. It can be used with a variety of root vegetables. It is not suggested for small-scale onion growers to cut off the green tips of bulb onions, as this practice considerably increases the risk of rot if the bulbs are not dried fast under controlled conditions. The green tops are mechanically removed before harvest in large commercial production, which frequently uses artificial heat and forced ventilation to dry its products. Small-scale production will not be able to afford this method. As long as the trays are well-ventilated, field-dried onions can be stored for up to two months under ambient circumstances. It is imperative that dried onions are kept away from wet soil.

Inhibition of Sprouting: When potatoes and onions are stored for up to eight months in temperate areas, sprouting is an issue. In warmer climates where growers may produce more than one crop a year, long-term storage may not be essential. Sprouting can be reduced using two methods:

»   Cultivars requiring longer durations of dormancy information on the storage properties of local types can be requested from seed and plant material suppliers.

»   The preservation of potatoes and onions through the application of chemical sprout suppressants. Before harvest, the crop needs to be treated with a variety of pesticides. Others, such as tecnazene, are sprinkled on top of the potatoes as they're loaded into the storage facility in powder or granule form. However, except for large-scale production and storage, suppressants should only be utilized after consulting with extension personnel. In the context of tropical root and tuber crops, nothing is known regarding the efficacy of sprout suppressants.

Fungicide Application

The use of post-harvest fungicides to prevent rot is common on a number of key crops that are either kept or transported long distances (citrus, bananas, apples, etc.). When packing produce that has been washed and dried, fungicides are typically employed.

Application method

Spray or mist: Mist or spray is both acceptable verbs. Mechanized spray rigs with moving belts or roller conveyors are employed for large-scale commercial operations, where the application is done by hand-held knapsack sprayers. To achieve comprehensive covering, produce is sprayed into the runoff.

Drenching: This is accomplished by pumping fungicide over the product that is being conveyed beneath it on a belt or roller conveyor in a cascade fashion. High flow rates through the pump keep the mixture stirred without the need for nozzles to wear out or become clogged. A non-foaming wetting agent may be added to the suspension to prevent the fungicide from dragging out if foaming occurs.

Dipping: The fungicide combination is prepared in a tiny container and the produce is dipped by hand in small batches. The excess fungicide drains back into the bath. In order for the fungicide to work properly, it must be shaken frequently. The use of rubber gloves is recommended for workers who dip their hands in fungicides, as they may have an allergic reaction.

Smoke or fumigant: The use of smoke or other irritants. Dust or vapour form of fungicide can be used in sealed containment chambers or bulk storage facilities. These kinds of procedures are extremely uncommon. Fumigation of bulk storage facilities necessitates specialized expertise, which is typically provided by outside contractors.

Hot water: The comforting warmth of running water. Even though hot-water dips have been proven to be successful in controlling post-harvest deterioration of some tropical fruits, the therapy has not been extensively adopted because of the difficulty in implementing it commercially. Anthracnose can be controlled with a hot fungicide dip that has been used commercially in Australia. A high degree of precision is required, and there is minimal room for error. Small-scale production is not a good fit for this method.

Controls on fungicide treatment

When using fungicides after harvest, they are typically subject to stricter regulations than when they are used while crops are growing. In terms of post-harvest treatment of fresh produce, there is a limited selection of chemicals available, with rigorous limits on concentrations and acceptable residue levels. When using a post-harvest fungicide, it's important to keep in mind that:

» Allowed to be used on the crop after harvest; beneficial in preventing post-harvest disease outbreaks on that crop acceptable.

» Utilized strictly in accordance with the directions provided by the manufacturer and at the acceptable concentrations.

» Continuously stirred to prevent sedimentation during usage.

» Workers applying fungicides should follow all safety procedures and wear appropriate protective clothes, according to those in charge of operations.

Prevetion of Losses During Storage

Controlled conditions: In the context of fresh produce, the term "storage" is now nearly universally understood to denote the storage of produce in a climate-controlled environment. Some significant crops, like potatoes, are stored on a huge scale to fulfill a regular demand and stabilize prices, but it also provides year-round access to a variety of local and exotic fruits and vegetables for people in industrialized countries and the affluent inhabitants of developing countries. However, storage in a controlled environment is not practicable in many underdeveloped nations where seasonal plant items are held back from sale and released gradually due to the expense and lack of infrastructure development, maintenance and administrative abilities. However, even in developed countries, many people still preserve and store fresh produce using traditional ways for their own consumption.

Storage Potential: There are just a few options for preserving even the most durable fresh produce when stored in ambient circumstances (i.e., food that is most susceptible to spoilage).

Organs of survival: Dormancy is a period of time following harvest when the edible components of various crops such as Irish potatoes, yams, beets, carrots, and onions are at their lowest nutritional value. If the right conditions are in place, this dormancy period can be extended to the maximum length conceivable. The storage potential is the name given to this property. It's crucial to know that different cultivars of the same crop have varying storage capacities. Local growers and seed providers are typically able to answer questions on this topic.

Edible reproductive parts: Leguminous plants' fruits and seeds are the primary source of these (peas and beans). Products in their fresh state have a short shelf life that can only be minimally prolonged by refrigerating them. Pulses can also be made by drying them and storing them in a jar. When stored properly, pulses have a lengthy shelf life and don't have the same storage issues as fresh food.

Fresh fruits and Vegetables: These include leafy green veggies, succulent fruits, and flower sections that have been changed (e.g. cauliflower, pineapple). Under normal conditions, their storage capacity is quite restricted. They decay quickly due to their high moisture content and rapid respiration rates, which induce rapid heat buildup and depletion of their high moisture content. Sun-drying or simple household processing into conserves and pickles are traditional methods of preservation. Even in ideal conditions, most fresh fruits and vegetables have a storage life of only a few days.

Factors affecting of Storage Life

Other biological and environmental variables severely reduce the natural post-harvest life of all sorts of fresh produce:

Temperature: As food stocks and water content dwindle, the rate of natural disintegration of all produce increases as temperatures rise. Cooling food will increase its shelf life by reducing the pace of disintegration.

Water loss: Water evaporation High temperatures and produce damage can significantly increase the loss of water from stored produce beyond what is necessarily lost due to natural reasons. Only store undamaged product at the lowest temperature tolerated by the crop for maximum storage life.

Mechanical harm: Damage produced during harvesting and subsequent handling accelerates the degradation of product and makes it vulnerable to decay microbes. Mechanical damage to root crops will result in significant losses due to bacterial decay, which must be addressed by curing the roots or tubers prior to storage.

Storage decay: The infection of mechanical damage is the leading cause of fresh produce decay during storage. Furthermore, decay microbes attack many fruits and vegetables, penetrating through natural openings or even through unbroken skin. These diseases can develop during the plant's growth in the field but remain latent until after harvest, typically becoming noticeable only during storage or ripening.

Ventilated stores: Stores that is well-ventilated. Naturally ventilated buildings can be utilized to store long-lasting vegetables such as roots and tubers, pumpkins, onions, and hard white cabbage. Such stores must be custom-designed and built for each planned location. Any sort of construction can be employed as long as it provides unrestricted air circulation through the structure and its contents. The following requirements must be met:

> » The structure should be placed in an area where low night temperatures occur during the desired storage duration.

> » It must be angled so that it makes the best advantage of the prevailing wind for ventilation.

> » The material used to cover the roof and walls should provide insulation from the sun's heat. The use of grass thatch on a bush-pole structure can be quite beneficial, especially when wetted to promote evaporative cooling.

> » If the expense allows, double-skinned walls will give higher insulation.

> » White paint applied to man-made surfaces will help to reflect the heat of the sun.

> » If trees do not interfere with the prevailing air flow, the structure should be built in their shade. Be wary of bushfires and trees that fall during storms.

> » Include ventilation gaps beneath the floor and between the walls and ceiling to allow for adequate air flow.

» Install movable louvers and adjust them to limit the flow of warm air entering the store during the day if the store is prone to cold night temperatures.

In warm regions, simple open-sided, naturally ventilated shelters can be utilized to store seed potatoes at high altitudes. If exposed to light for more than a few hours, table potatoes will turn green, develop a harsh flavour and even become hazardous.

Clamps: In Europe and Latin America, they are simple, low-cost buildings used to store root crops, mainly potatoes. In warm areas, the potatoes are planted on a straw bed 1 to 3 m wide, but no more than 1.5 m wide. Along the bottom, a ventilating duct should be installed. The heaped potatoes are covered with around 20 cm of compressed straw, which can then be encased in soil up to 30 cm deep, applied without compaction. The clamp mechanism is adaptable to diverse climate conditions. Extra straw casing may be used instead of soil in warm climates to provide additional ventilation.

Other simple storage methods: Windbreaks are narrow, wire-mesh, basket-like constructions about 1 m wide and 2 m high of any practical length, on a raised wooden foundation, and are used in the field for short-term storage of dried onions. The onions are topped with a 30 cm covering of straw that is held in place by a polythene sheet connected to the wire mesh. To achieve optimal drying and ventilation, the windbreak is built at right angles to the prevailing wind. Onions can also be woven into plaits on twine and stored in a cold, dry spot for several months.

***Refrigerated and controlled-atmosphere* storage:** Refrigerated storage may be utilized in a cold-chain operation to transport regular consignments from production areas to urban markets and retailers for large-scale commercial operations. This can be a very complex procedure that demands skilled organization and administration. Cold storage can also be used to store seasonal crops such as potatoes and onions for an extended period of time. Some fruits, such as apples, can have their storage life extended by combining refrigeration with a controlled environment composed of oxygen and carbon dioxide. These are costly enterprises with significant maintenance and operating costs that necessitate competent and experienced management. They are mostly used in small-scale production in poor countries. Many perishable food crops are produced seasonally in most nations, making them available only for brief periods of the year. They are produced in bigger quantities than the market can consume during this brief period, thus the surplus of many of these crops must be processed and conserved to avoid food waste and money loss to the grower. In industrialized countries, modern methods of food storage and preservation, such as refrigeration and freezing, are now commonly used.

These methods are, however, uncommon in many poor nations, yet surpluses of many seasonal indigenous products can be conserved for later consumption using a variety of simple and inexpensive processing processes.

Chapter - 11

Advance Technnologies
for Food Preservation

The desire from consumers for meals that promote health and have high nutritional and nutraceutical benefits has led to novel food processing methods and contemporary trends in the industry. The goal of the food industry has been to create safe food with a long shelf life from the beginning of time. However, today, consumers seek foods that are also highly nutritious, include bioactive substances, and have pleasing sensory qualities. Different food preservation techniques target microorganisms because they are the primary target species for food decomposition and poisoning. Industry-wide food processing techniques rely on either microbial inactivation or microbial growth inhibition. The flavour, nutritional content, and appearance of food can be negatively impacted by conventional heat-dependent pathogen-reduction techniques including thermization, pasteurisation, and in-container sterilising. Due to the growing customer desire for better and higher quality food items targeted specifically at them, alternative methods to standard thermal processing of food have drawn a lot of attention. These methods are referred to as "novel" or "emerging" approaches. Consumer desire for fresh, high-quality, healthful products that are devoid of chemical preservatives but still safe have led to a need for revolutionary processing technologies in the food sector. Although technically challenging, the trend toward using "natural" ingredients (colours, flavours, or preservatives) has prompted the need for research into gentler, more energy-efficient, but equally effective processing technologies that can preserve the structure and, consequently, function and benefits of novel ingredients while also maintaining the nutritional and other qualities of the food product. The primary objective of food and beverage businesses has always been to improve product quality. Here, a few of these cutting-edge technologies are discussed.

Food Irradiation

Food and food packaging are subjected to ionising radiation, such as that from gamma rays, x-rays, or electron beams, during the food irradiation process. By successfully eliminating the germs that cause food spoilage and food-borne illness, food irradiation enhances food safety, prolongs product shelf life (preservation), prevents sprouting or ripening, and serves as a tool for pest control of invading insects and other pests. Food can be irradiated dry, moist, frozen, or thawed. It is also known as "Cold Sterilization" because the method doesn't generate any heat. Food can be sterilised or pasteurised without affecting its texture or freshness, and due to the method' efficiency, even pre-packaged meals can be treated with it.

Food is physically treated by being subjected to ionising radiation, a process known as food irradiation. high-energy radiation used to ionise molecules and remove electrons from atoms By killing or inactivating contaminants in food, such as insects, mould, yeast, and bacteria that cause food to decay, irradiation extends the shelf life of products. It also delays the ripening of fruits and vegetables and prevents unintended potato and other similar product sprouting.

Foods are safe as they are itself never comes in contact with radioactive materials.

Table 1. Application of ionising radiation in various does

Low dose		0.03-0.15 kGy	Sprout inhibitor
	≤ 1 kGy	0.25-0.75 kGy	Ripening inhibitor
		0.25-1 kGy	Insect disinfestation
Moderate dose		1.5-3.0	Reducing of spoilage microbes to improve the shelf life of sea food under refrigeration
	1-10 kGy	3.0-7.0 kGy	Elimination of microbes in fresh and frozen meat
		10 kGy	Removal of microorganism from spices
High dose	≥ 10 kGy	25-70 kGy	Sterilization of food products

Mechanism

The main mechanism by which the micro-organism is killed are-

» Impact on microbial DNA and RNA molecules. Because the DNA double helix cannot unwind, microorganisms are unable to reproduce.

» Causing a chemical connection to be destroyed, disrupting the internal metabolism of the cell.

» Damage from radiation is caused by free radicals reacting with cellular DNA.

Irradiation can serve many purposes

» *Prevention of Food-borne Illness*: to effectively eliminate organisms that cause food-borne illness, such as *Salmonella* and *Escherichia coli* (*E. coli*).

» *Preservation*: to extend the shelf life of food by eliminating or inactivating organisms that lead to spoilage and decomposition.

» *Control of Insects*: to eradicate insects found in or on imported tropical fruits. Additionally, other pest-control methods that might harm the fruit are less necessary after irradiation.

» *Delay of Sprouting and Ripening*: to inhibit sprouting (e.g., potatoes) and delay ripening of fruit to increase longevity.

» *Sterilization*: foods can be sterilized with irradiation and then kept for years without refrigeration. For patients with extremely compromised immune systems, such as those with AIDS or receiving chemotherapy, sterilized foods are helpful in hospitals. Foods that are sterilized using irradiation are subjected to far more treatment than those that are deemed safe for general consumption.

Source of radiation:

There are three sources of radiation approved for use on foods.

» Gamma rays are emitted from radioactive forms of the element cobalt (Cobalt 60) or of the element cesium (Cesium 137). Gamma radiation is used routinely to sterilize medical, dental, and household products and is also used for the radiation treatment of cancer.

» X-rays are produced by reflecting a high-energy stream of electrons off a target substance (usually one of the heavy metals) into food. X-rays are also widely used in medicine and industry to produce images of internal structures.

» Electron beam (or e-beam) is similar to X-rays and is a stream of high-energy electrons propelled from an electron accelerator into food.

Table 2. Application of radiation on fruits and vegetables

Produce	Effect	Dose (kGy)
Apple	Control scald and brown core	1.5
Apricot, peach	Inhibit brown rot	2
Banana	Inhibit ripening	0.3-0.35
Mushroom	Inhibits stem growth and cap opening	2
Strawberry	Inhibits grey mould	2
Papaya	Disinfection of fruit fly	0.25
Tomato	Inhibits *Alternaria rot*	3

Detection of food irradiation

Food after irradiation doesn't undergo any significant physical, chemical, or sensory alterations. The detection of irradiation becomes significant when administered at or below the commercially authorized levels and concentrates on tiny changes in the chemical composition, physical characteristics, or biological characteristics of the food. A few techniques for radiation detection include:

Thermo-luminescence detection of irradiated food

Food-contaminating silicate minerals store energy through charge-trapping reactions as a result of being exposed to ionising radiation. The release of this energy results in quantifiable thermo-luminescence (TL) light curves when isolated silicate crystals are heated under controlled conditions. The majority of the time, a density separation stage is used to separate silicate minerals from the food. A first glow of the separated mineral extracts is recorded (Glow 1). Since various amounts and/or types of minerals (quartz, feldspar etc.) exhibit very variable integrated TL intensities after irradiation, a second TL glow (Glow 2) of the same sample after exposure to a fixed dose of radiation is necessary to normalize the TL response. The TL glow ratio, thus obtained, is used to indicate radiation treatment of the food, since irradiated samples yield higher TL glow ratios than that of unirradiated samples. Detection of irradiated herbs, spices (6 kGy), irradiated shellfish (0.5 kGy to 2.5 kGy), irradiated fresh and dehydrated fruits and vegetables (1 kGy).

ESR (Electron Spin Resonance Spectroscopy)

When exposed to radiation, distinct mono- or disaccharides may predominate in the sample, resulting in diverse ESR spectra. Irradiation won't result in any distinctive

ESR signals if the sample doesn't include any sugar crystals. It has been confirmed that raisins, dried papayas, dried mangos, and dried figs were all irradiated. The crystalline of the sugar in the sample has a significant impact on the lower limit of detection. This method's applicability depends on the sample containing enough crystalline sugar at every stage of handling between irradiation and testing. If necessary, irradiating a piece of the sample and retesting can be used to confirm the sensitivity to radiation. Prior to testing, it is crucial that dried fruits have not been rehydrated.

Photo Stimulated Luminescence

On most foods, there are mineral fragments that are typically silicates or bioinorganic minerals like hydroxyl-apatite from bones or teeth or calcite from exoskeletons or shells. When subjected to ionising radiation, these materials store energy in charge carriers that become stuck at structural, interstitial, or impurity locations. The process consists of screening (first) PSL measurements to establish the status of the sample and a second measurement, which is optional, after a calibration radiation dosage to ascertain the sample's PSL sensitivity. In theory, any food containing mineral debris can be irradiated, and the PSL method can be used to detect it. Signals that are below the lower threshold (T1) are typically found in naturally occurring materials, though they can also come from low sensitivity irradiation materials. It may be possible to tell these circumstances apart using calibration. To prevent false negative results, it is generally advised to use calibrated PSL measurements on "clean" spices like nutmeg and ground white and black pepper as well as shellfish with low mineral contents. Unblendable products yield the best results.

Aerobic Plate Count (APC) Screening method

Utilizing the aerobic plate count approach, a microbiological screening technique can be used to identify herbs and spices that have undergone radiation therapy. A limitation of the method occurs when there are too few microbes in the sample (APC 103 cfu/g), so it is advised to confirm positive results using a standardised method (e.g. EN 1788, prEN 13751) to specifically prove an irradiation treatment of the suspected food. The APC technique is not radiation specific. The DEFT/APC difference of counts can be comparable to the difference of counts obtained after irradiation if fumigation or heat treatment had been utilised for decontamination. However, it is possible to spot the application of fumigation. Some spices such as cloves, cinnamon, garlic and mustards contain inhibitory components with an anti-microbial activity which may lead to decreasing APC.

Microbiological screening method

In this method comprising two procedures, which are carried out in parallel. It permits the identification of an unusual microbiological profile in poultry meat, for example. The presence of a large excess population of dead micro-organisms can under certain circumstances be presumptive of irradiation treatment, which means, that the results of the procedure of the determination of endo-toxin concentration in the test sample using the Limulus Amoebocyte Lysate (LAL) test and of the procedure of the enumeration of total Gram negative bacteria (GNB) in the test sample are not radiation specific. This method can give only an indication of a possible treatment by ionizing radiation. The method is of particular use to routine microbiological laboratories, which may be involved in the examination of foods.

Cold Plasma Technique

The novelty of cold plasma (CP), among other cutting-edge non-thermal technologies, lies in its non-thermal, affordable, adaptable, and environmentally friendly characteristics. Cold plasma (CP) is a relatively new technology that has emerged as an alternative source for surface sterilization and disinfection, for ensuring the quality and safety of minimally processed food. A quasi-neutral ionised gas known as plasma is made up mostly of electrons, ions, and reactive neutral species in their fundamental or excited states. Based on the thermal equilibrium, there are two plasma classes—denominated non-thermal plasma (NTP) or cold plasma and thermal plasma. Cold plasma is generated at 30-60°C under atmospheric or reduced pressure (vacuum), requires less power, exhibits electron temperatures much higher than the corresponding gas (macroscopic temperature), and does not present a local thermodynamic equilibrium. The cold plasma approach was initially used in the polymer, biomedical, and surface engineering industries to increase antibacterial activity. Cold plasma has received a lot of interest for the non-thermal preservation of agricultural products due to its outstanding antibacterial properties. The ions and uncharged molecules only gain a small amount of energy and maintain a low temperature, making it suitable for food products that are sensitive to heat.

Applications of cold plasma (CP)

Cold plasma is an unique non-thermal technology that was formerly utilized in the biomedical technology industry to sterilize thermo-labile materials. It is now being applied to the food industry. Current cold plasma research is concentrated on its uses for food decontamination, enzyme inactivation, toxin breakdown, waste water treatment, and packaging improvements, particularly in the food industry. Cold plasma has been shown to be effective for inactivating food-borne pathogens and spoilage bacteria in the context of food preparation. Recently, different inactivation mechanisms for Gram positive and Gram negative bacteria by cold plasma. They observed that cold plasma inactivation of Gram positive bacteria (*Staphylococcus aureus*) was mainly due to intracellular damage and little envelope damage whereas Gram negative bacteria (*Escherichia coli*) was inactivated mainly by cell leakage and low-level DNA damage. The potential of cold plasma to degrade numerous food contaminants, particularly mycotoxins, has emerged in recent years, piqueing the interest of the food industry. Plasma treatment is used for surface cleansing, surface sterilization, and surface treatments like cleaning, coating, printing, painting, and adhesive bonding in the case of packing materials.

Principle cold plasma technique

» At atmospheric pressure, a process gas is passed via an electric field to produce cold plasma.

» Impact ionisation procedures are started by accelerated electrons from ionising processes.

» Free e⁻ colliding with gas atoms transfer their energy, thus generating highly reactive species that can interact with the food surface.

» The e⁻ energy is sufficient to dissociate covalent bond in organic molecules.

Advantage of cold plasma technique

» It is eco-friendly technology.

» Low temperature (below 50°C) cold plasma affords great microbial activation efficiency.

» It's compatible with most existing packaging and modified atmosphere.

» Cold plasma does not leave any residues.

» It is an energy efficient technique.

Limitation of cold plasma technique

» Lipid oxidation, colour changes, a loss of hardness, and an increase in acidity are possible symptoms.

» High voltage plasma generation necessitates additional safety precautions.

» Cost of cold plasma generation is very high.

» Because reactive oxygen species cause oxidation, treating high-fat foods with cold plasma, an oxygen-containing gas mixture, may not be appropriate.

High Pressure Processing (HPP)

The trend in food consumption has recently changed from typical thermally processed items to minimally processed foods that are natural, healthy, microbiologically safe, shelf-stable, and have natural texture. This is due to the growing consumer awareness of these issues. High pressure processing (HPP) is one such tested technology that has been gaining popularity in the food processing industry for the past few decades and is addressing all of these customer demands.

High pressure processing (HPP), also known as cold pasteurization, is the most popular non-thermal technology because it has the ability to increase food shelf life by eliminating microorganisms. Beyond the novelty and freshness of the product, HPP can increase the efficiency of processes like dehydration, blanching, rehydration, osmotic dehydration (OD), freezing, and thawing. However, its main drawback is the high initial capital expenditure, which may be improved by a decrease in operating costs due to HPP's lower energy requirement. The HPP has been marked as a substitute to pasteurization by United States National Advisory Committee (USNAC) on Microbiological Criteria for Foods. The other food safety certification agencies, such as the United States Food and Drug Administration (USFDA) and the United States Department of Agriculture (USDA), have approved the applications of HPP in food. Because HPP can kill pathogens, bacteria that cause food deterioration and enzymes at a very low temperature without damaging other aspects of the food, it is advised that it be used in the preparation of foods with the required textural and nutritional attributes.

Principle of HPP:

The HPP is based on two basic principles

1. Le Chatelier's principle

Pressure enhances any phenomenon (phase transition, change in molecular configuration, chemical reaction) accompanied by a reduction in volume. Pressure consequently causes the systems to change to the lowest volume system.

2. Isostatic principle (Pascal's Law)

Food products are compressed by constant pressure coming from all sides, and when the pressure is released, they regain their original shape. Because pressure transfer to the core is not affected by mass or time, the products are crushed regardless of their size and geometry, which minimises the process. If a food product is being examined at microscopic levels, as long as pressure is being exerted consistently throughout.

Application of HPP in food processing:

Processing of fruits, vegetable and their products

One of the more promising technologies for processing fruits, vegetables, and their products is HPP (Table-3). High-acid fruits are typically the most popular options in HPP operations because they decrease nutrient quality losses during leaching and make high pressure blanching of fruits and vegetables more widely used. HPP is also highly fruitful in osmotic dehydration that activates the more open food tissue thus improving the mass transfer. The orange juice treated with HPP has a SL of 21 days and is microbiologically safe, retaining organoleptic and nutritional properties alike to fresh juice. Moreover, HPP has potential in the pasteurization of fruit juices, vegetable mixes, and RTE meals so as to increase their SL of 1–2 months at 4°C.

Table 3. Application of high pressure processing in fruits, vegetable, and their products

Foods and Food materials	Optimum Experimental Condition(s)		
	Pressure (MPa)	Holding time (Min)	Temperature (°C)
Fruit and vegetables juice	400–800	10	25–50
Apple juice	400	10	–
Broccoli tomatoes and carrot (crushed wateror liquid extracts)	500–800	–	25 or 75
Fresh cut pineapple	340	15	–
Guava puree	600	15	25
Kiwifruit, melon,pears, and peaches	400	30	5 or 20

Lemon juice	450	2, 5 or 10	-
Asparagus, cauliflower, lettuce, onion, spinach, and tomato	300–400	-	-
Litchi	200–600	10 or 20	20-60
Orange Juice	350	1	30
Raspberry puree	200–800	15	18-22
Strawberry jam	400	5	Room temperature

Advantage of high pressure processing

» Nutrient, colour, flavour and other quality qualities are not significantly harmed by the process because processing is carried out at ambient or even cold temperature.

» Reduce the processing time because there are just two cycles of pasteurization and de-pasteurization rather than heating and cooling periods.

» Application of all time of foods, whether liquid or solid.

» There is very little chance of post-harvest contamination.

» Little chance of any toxicity.

Limitation

» Complex material handling.

» Little effect of food enzyme activity.

» Expensive machinery.

» Small potential for continuous processing.

Ohmic heating:

In the literature, ohmic heating—also referred to as Joule heating, electro conductive heating, electrical resistance heating, direct electrical resistance heating, and electro heating—is a thermal-electrical technique in which food is in contact with the electrodes. A crucial factor in the design of a successful ohmic heater is electrical conductivity. Ohmic heating has a wide range of potential uses, such as food heating, pasteurization, sterilization, evaporation, dehydration, fermentation, and blanching. In addition to heating, an applied electric field during ohmic heating results in electroporation of cell membranes, which raises extraction rates and lowers the temperature and enthalpy of gelatinization. Ohmic heating causes food to heat up more quickly while also maintaining its colour and nutritional content. The use of ohmic heating affects the water absorption index, water solubility index, thermal characteristics, and pasting properties. Pre-gelatinized starches produced by ohmic heating lower processing energy requirements. But its higher initial cost, lack of its applications in foods containing fats and oils, and less awareness limit its use.

Principle of Ohmic Heating

The principles of ohmic heating are very simple. The principle of ohmic heating is the generation of heat by the passage of alternating electrical current (AC) through an electrically resistive body, such as a liquid-particulate food system. The electrodes at the product body's two ends receive AC voltage. The rate of heating is directly proportional to the square of the electric field strength, E, and the electrical conductivity. By altering the electrode spacing or the applied voltage, the electric field intensity can be changed. The product's electrical conductivity and its temperature sensitivity, however, are the most crucial elements. The electrical conductivity of all the phases must be taken into account if the product has more than one phase, such as when a mixture of liquid and particles is present. The electrical conductivity increases with rising temperature, suggesting that ohmic heating becomes more effective as temperature increases, which could theoretically result in runaway heating. The heating characteristics of the system might become quite complex due to a variation in electrical resistance and its temperature dependency between the two phases. Since ionic content affects electrical conductivity, it is possible to modify the product's electrical conductivity (in both phases) using ion (e.g., salt) levels to obtain efficient ohmic heating.

Type of ohmic heating

Flange heating and screw plug heating are the two most widely utilised heating techniques in the food sector, which both are powered by electricity. Their capacity to provide consistent heating across all products may be their best quality.

Ohmic heating process

The process of continuous flow ohmic heating is shown schematically as ohmic heating.

A feed pump hopper is used to introduce a viscous food product with particulates into the continuous-flow ohmic heating system. The product is subsequently heated to process temperature as it passes through a set of electrodes in the ohmic column. The product then spends a specific amount of time in the holding tubes to establish commercial sterility. The product then passes through tubular coolers and is stored in storage tanks until it is filled and packaged. The heater assembly, power supply, and control panel are the three parts that make up the majority of ohmic heating system setups.

Application of ohmic heat

Due to their thermal conductivity and resistance characteristics, salt, water, and fat contents affect how well electricity is converted to heat. Due to their greater electrical resistance and matched conductivity, food particles heat up more quickly than the liquid matrix in particulate foods, which can result in more even heating. This keeps the liquid matrix from overheating while still digesting the particles with enough heat. To illustrate the impact of composition and salt concentration. Table-4 provides electrical conductivity values for a selection of foods. High electrical conductivity values signify a greater concentration of ionic compounds suspended in the final product, which is inversely related to the rate of heating. This value rises in the presence of polar substances like acids and salts while falling in the presence of non-polar substances like lipids. Electrical conductivity of food materials generally increases with temperature, and can change if there are structural changes caused during heating such as gelatinization of starch. Density, pH, and specific heat of various components in a food matrix can also influence heating rate.

Table 4. Electrical conductivity of selected food

Food	Electrical conductivity (S/m)	Temperature (°C)
Apple juice	0.239	20
Carrot	0.041	19
Carrot juice	1.147	22
Coffee (black)	0.182	22
Coffee (black with sugar)	0.185	22
Coffee (with milk)	0.357	22
Chicken meat	0.19	20
Beer	0.143	22

Advantage

» Heating food materials internally eliminates some of the non-uniformity typically associated with microwave heating due to limited dielectric penetration and the restrictions of conventional heat transfer.

» The product heats up volumetrically and does not encounter a significant internal temperature differential.

» It is possible to reach higher temperatures in solids than in liquids, which is not possible with conventional heating.

» Reducing the likelihood of food product burning and fouling on heat transfer surfaces, which minimizes mechanical damage and improves nutrient and vitamin retention.

» High energy efficiency because 90% of the electrical energy is converted into heat.

» Optimization of capital investment and product safety as a result of high solids loading capacity.

» Ease of process control with instant switch-on and shut-down. Reducing maintenance cost (no moving parts).

» Ambient-temperature storage and distribution when combined with an aseptic filling system.

» A quiet environmentally friendly system.

Ultrasound

Food processing is today a highly interdisciplinary subject and is no longer simple or straightforward. New methods have emerged to increase shelf life, reduce danger, safeguard the environment, and enhance functional, sensory, and nutritive qualities.

Ultrasound is a type of environmentally friendly technology with numerous uses in food processing, preservation, and quality assurance. It is a new technology that alters several aspects of food goods. Application of mechanical waves with a frequency above the range of human hearing (more than 16 kHz) constitutes ultrasound technology. The ultrasonic spectrum can be further broken down into low frequency (20-100 kHz), high power (>100 kHz), and low power (1 W/cm2) ultrasound depending on frequency and intensity.

During the processes of freezing and crystallization, defrosting and thawing, drying, meat tenderization, pickling, and emulsification and homogenization, different frequencies of ultrasound can be effectively used to change the physiochemical properties of food. The method can also be used to sterilize, defoam, depolymerize, and inactivate enzymes in addition to extracting bioactive components. Ultrasound is a low-cost technique that has little impact on food quality; its sole drawback is that it can't be used on some meals since they contain tiny air bubbles.

Mechanisms

The cavitations phenomenon, which results from the development of bubbles or cavities in liquid, has been attributed as the cause of ultrasound's fatal effect. The collapse of the bubble results in a strong localized change in temperature (4700°C) and pressure (30,000 psi), which is fatal to microorganisms and more likely to cause shear disruption, cavitations, membrane thinning, etc. Gram -ve bacteria are more vulnerable than Gram +ve bacteria.

Methods of ultrasound

By increasing the effectiveness of inactivation, ultrasound can be utilized for food preservation in conjunction with other therapies. Numerous investigations have combined either pressure, temperature, or both pressure and temperature with ultrasound.

Ultrasonication: The use of ultrasonic energy at low temperatures is known as ultrasonication. It can therefore be utilized for products that are sensitive to heat. However, stable enzymes and/or bacteria that can take a lot of energy to inactivate call for extended treatment times. Depending on the ultrasound strength and time of application, there may be a rise in temperature during ultrasound application; this needs to be controlled to optimize the process.

Thermosonication: Thermosonication combines the use of heat and ultrasound. The product is concurrently exposed to mild heat and ultrasonic waves. Compared to using just heat, this technique has a stronger impact on the inactivation of microorganisms. Lower process temperatures and processing times are needed to attain the same lethality levels as with conventional methods when thermosonication is employed for pasteurization or sterilizing.

Manosonication: It is a combination technique that uses both pressure and ultrasound. By mixing ultrasound with moderate pressures and low temperatures, manosonication offers to inactivate enzymes and/or bacteria. At the same temperature, it has a greater inactivation efficiency than only ultrasonic.

Manothermosonication: Manothermosonication combines pressure, heat, and ultrasound. When compared to thermal treatments at the same temperatures, MTS treatments inactivate a number of enzymes at lower temperatures and/or faster times. The level of inactivation is increased by the media's cavitation or bubble implosion, which is maximized by the applied temperature and pressure. Mantheromonication can inactivate microorganisms with exceptional thermotolerance. Manothermosonication can also inactivate several thermo resistant enzymes, including Pseudomonas heat-labile lipases and proteases as well as polyphenoloxidase, peroxidase, and lipoxygenase.

Application

For microbial inactivation, freezing, drying, and extraction, ultrasound has been used on meat products, vegetables and fruits, cereal goods, aerated meals, honey, food gels, food proteins, and food enzymes.

Table 5. Some applications of US that are of considerable interest in the food industry

Food	Purpose
Cheese	Reduction of product losses (cut)
Potatoes	Reduction of structural damage (freezing)
Food systems	Time saving (marinating, filtration, and oxidation)
Food materials	Reduction of heating and cooling costs (cooking and freezing)
Skim milk and sunflower oil	Shelf-life improvement (emulsification)
Food and agricultural products	Improvement of product structure (mixing and drying)
Fruits and vegetables	Microbiological safety (anti-foaming, degassing, and sterilization)

Advantage

» In accordance with its strength.

» Enzyme activation or inactivation, mixing, and homogenization are all accomplished using ultrasound.

» Emulsification, dispersion, preservation, stabilization, dissolution and crystallization, hydrogenation, tenderization of meat, ripening, ageing and oxidation and as an adjuvant for solid-liquid.

Limitation

Because microorganisms and enzymes are so resistant to ultrasound therapy, the strength of the treatment may adversely affect the texture and other textural aspects of food and significantly lower its sensory quality.

Conclusion

These unique, novel methods of food preservation are new approaches with a wide range of uses in the food sector. The food that needs to be processed, the type of microorganisms present, and their load all need to be considered while building these applications. Despite some drawbacks, if used in conjunction with other food processing methods, these technologies can improve food preservation with little negative impact on food quality.

References

Ahvenainen R. Novel Food Packaging Techniques. Cambridge UK: *Wood head Publishing*. 2003; ISBN 978-1-85573-675-7.

Bagchi, A. Intelligent sensing and packaging of foods for enhancement of shelf life: concepts and applications. *International Journal of Scientific & Engineering Research*. 2012; 3(10): 1–13.

Coles, R.; Mcdowell, D. and Kirwan, M. J. Food Packaging Technology. Oxford, UK: *Blackwell Publishing*. 2003; ISBN 978-0849397882.

Fung, F.; Wang, H. and Menon, S. Food safety in the 21st century. *Biomed. J.* 2018; (77): 347-355.

Girdhari, L.; Siddappa, G. S. and Tondon, G. L. Preservations of fruits and vegetables. *Publication and Information Division, Indian Council of Agricultural Research, New Delhi*. 1986.

IAEA (International Atomic Energy Agency). Irradiation to ensure the safety and quality of prepared meals, *Vienna, Austria*. 2009.

Kumar, V. and Kamra, S. Ready Reckoner for Pomology. *New Vishal Publication, New Delhi*. 2018.

Kuswandi, b.; Wicaksono, y.; Jayus, Abdullah, A.; Lee, Y. H. and Ahmad, M. Smart packaging: sensors for monitoring of food quality and safety. *Sensory and Instrumentation for Food Quality*. 2011; Vol. 5, p. 137–146.

Mills, A. Oxygen indicators and intelligent inks for packaging food. Chemical Society Reviews. 2005; 34: 1003–11.

Nayak, S. L. and Prajapati U. Insights to Processing of Horticultural crops. *New Vishal Publication, New Delhi*. 2020.

Nummer, B. A. Historical Origins of Food Preservation. *National Center for Home Food Preservation. National Center for Home Food Preservation, New Delhi*. 2016.

Pandit, P. S.; Ram B. and Ahmad, T. Post Harvest Technology & Processing of Horticultural crops (at a glance). *New India Publishing Agency, New Delhi*. 2014.

Rathor, R. S.; Mathur, G. S. and Chasta, S. S. Post harvest management and Processing of Fruits and Vegetables. *Directorate of knowledge Management in Agriculture, Indian Council of Agricultural Research, New Delhi*. 2012.

Scarano, S.; Mariani, S. and Minunni, M. Label free Affinity sensing: application to food analysis. *ACTA IMEKO*. 2016; (5): 36–44c.

Srivastava, R. P. and Kumar, S. Fruit and Vegetable Preservation: Principle and Practices, *CBS Publication and Distribution Pvt. Ltd*. 2002.

Thorne and Stuart. The History of Food Preservation. *Kirby Lonsdale, Cumbria, England: Parthenon*. 1986.

Vaillant, R. Method, system and component for controlling the preservation of a product. *United State Patent Publication*. 2010; US 7691634.

Weng, X. and Neethirajan, S. Ensuring food safety: Quality monitoring using microfluidics. *Trends Food Sci. Technol*. 2017; (65): 10–22.

Printed in the United States
by Baker & Taylor Publisher Services